ACTION

THE TRUE CAUSE OF
GRAVITY REVEALED
By Jeff Lee

GRAVITY

Archway Publishing books may be ordered through booksellers or by contacting:

Archway Publishing
1663 Liberty Drive
Bloomington, IN 47403
www.archwaypublishing.com
844-669-3957

CENTER FOR REALITY PHYSICS
MindSight Publications

WEBSITE:

Centerforrealityphysics.com

Because of the dynamic nature of the Internet, any web addresses or links contained in this book may have changed since publication and may no longer be valid. The views expressed in this work are solely those of the author and do not necessarily reflect the views of the publisher, and the publisher hereby disclaims any responsibility for them.

Any people depicted in stock imagery provided by Getty Images are models, and such images are being used for illustrative purposes only. Certain stock imagery © Getty Images.

Interior Image Credit: MindSight Publications

ISBN: 978-1-6657-0184-6 (sc)
ISBN: 978-1-6657-0185-3 (e)

Library of Congress Control Number: 2021901095

Print information available on the last page.

Archway Publishing rev. date: 07/13/2021

DEDICATED TO MY PARENTS

S
MIND
G
H
T
PUBLICATIONS

What is Reality Physics?

The study of the fundamental nature of objective physical reality; why it is here, why it keeps on going on from one moment to the next, and why we are here stuck right in the middle of it, is the greatest challenge humankind has ever faced. Theoretical Physics Academia proposes a "static" four-dimensional configuration of space, time, mass and energy in their feeble attempt to reveal their: "Theory of Everything", which, if you really think about it, is really a complete "oxymoron".

I mean, if you are going to navigate the Solar System, do you want a "theory" to guide you, or do you want the correct explanation of how the universe really works to show you exactly where to go? This, oxymoronic condition has completely "hi-jacked" Theoretical Physics Academia and is the reason Reality Physics was invented.

With Reality Physics there is no theory, speculation, or hypothesis, there are only facts. If it can't be proven to be true, then it is not Reality Physics. This now allows us to leave the mythological realm of Theoretical Physics Academia behind and enter into a new type of universe where we now have three dimensions of space, and two (active) dimensions of time, that expand outward into space at the speed of light.

By projecting that mass causes a resistance to this two-dimensional, outward motion of time with space, we create an implosive motion of this inertial reference background now enabling us to reveal the actual cause of gravity. As we can immediately observe here, this is completely impossible with the four-dimensional, "static" universe of General Relativity, and can only be accomplished with this new "active" universe of Reality Physics. Reality Physics is the future and will open the door into a brand-new world of understanding and amazement that will now enable us to answer many fundamental questions we can only wonder about and contemplate today.

Gravity's Secret?

The one thing that all of us know is that the discovery of the true cause of gravity is indeed the "Holy Grail" of all scientific revelation.

Why? Because by knowing the true cause of gravity we all now enter into a completely new realm of scientific knowledge where we are finally able to understand the cause of inertia, the true origin of force, the real reason as to why matter has mass, and the connection between centrifugal force and gravity.

The secret is: The still position located within the inertial reference background of the universe, that exists as a plenum structure within three-dimensional space, can physically <u>move</u>, or <u>accelerate</u>, in a downward direction with respect to the still position of three-dimensional space <u>at the surface of a large object like a planet, or star.</u>

So, what could possibly cause this to happen? It is the ability of mass to cause a resistance to the active, two-dimensional, omnipositional, omnidirectional, outward motion of Time with Space at the speed of light" "**c**", or: **299,792,458 meters per second** at the object's surface

ACTION GRAVITY

THE TRUE CAUSE OF GRAVITY REVEALED
by JEFF LEE CENTER FOR REALITY PHYSICS

This presentation, for the first time ever in the history of physics, now reveals the true cause of gravity. This is accomplished by completely discarding the incorrect assumption of General Relativity that gravity is "curved space-time". By introducing a new type of physics called: REALITY PHYSICS we are able to show how gravitational acceleration results directly from the ability of mass to cause a resistance to the two-dimensional outward motion of time with space. This resistance of mass to the outward motion of time now produces a vicinity of less active pressure density at that location of the given body of mass, as compared to the greater pressure density existing throughout the rest of the universe. This now produces an implosive motion of this inertial background reference causing still position of the inertial background to implode down through space at the surface of the body of mass.

"GRAVITY!" It is the unrelenting force to which everything in the universe is constantly subjected. What really is gravity? What causes it? What is the actual mechanism that creates gravitational force? Why is gravity an acceleration? Why do all small objects (neglecting any wind resistance), no matter how heavy, always fall at the exact same rate? What is the relationship of gravity with acceleration, force and inertia? How is it that time passes slower in a gravitational field than it does in empty space? Why does time completely stop in a black hole? And, how is it that gravity can bend light?

The true cause of gravity is still one of the most perplexing and elusive mysteries in Theoretical Physics Academia today. It is interesting to note here that Physics academia has absolutely no idea what so ever as to where to even start to explain the true cause of gravity. Why? Because they are all in the **WRONG** universe. This means they could try forever to use General Relativity to explain gravity and they would find themselves no closer to the answer than they are right now.

It was Galileo who first put us on the right track as far as exposing the real truth revealing the fundamental nature of gravity when he discovered that: ***all solid objects (neglecting any wind resistance), regardless of their weight, fall at the same rate.*** Newton was then able to build upon this and other similar ideas as brought forth by Copernicus, Kepler and several other early scientists now enabling him to mathematically incorporate his equations describing the gravitational attraction of one object for another into his system of Classical Dynamics we are so familiar with today.

According to Newton, gravity was described as a force originating from within the inherent nature of the background reference of the universe itself and manifested itself as the physical attraction of one body of mass towards all of the others. Although Newton never really formally claimed to know the actual cause of gravity, he did postulate that gravitational force may be the result of: ***"varying differences in the density of the aether",*** which was a three-dimensional "plenum" type of structure that existed within the three-dimensional "absolute" space of his universe.

Small objects would then naturally move away from the vicinities of space located far away from large bodies of mass, where the aether was denser, to the vicinities of space, closer to the large bodies of mass, where the aether was less dense. This condition cannot exist in a "static" universe like the one as described by the General Theory of Relativity, but only in this new "active" universe as described here by the ACTION GRAVITY of Reality Physics.

As to what caused this aether to be denser far out in space and yet less dense near large bodies of mass, Newton never really gave any answer at all. However, for anyone to even contemplate this more than three centuries ago shows the extreme brilliance of Newton. He was also the first to extend the effects of gravity, as experienced down here on Earth, out into the Solar System, and beyond. This is why Newton is known today throughout the entire world as being the greatest contributor to have ever investigated the true nature of gravity. The equations he developed have proven to be correct and have lasted for over three and a half centuries now.

Fig.1 | **6 CLUES WE USE TO REVEAL THE CAUSE OF GRAVITY**

1. GRAVITY IS A CONSTANT ACCELERATION.

2. TIME SLOWS DOWN IN A GRAVITATIONAL FIELD.

3. ALL OBJECTS FALL AT THE SAME RATE.

4. LIGHT BENDS AROUND MASSIVE OBJECTS.

5. OBJECTS STILL IN SPACE EXPERIENCE NO FORCE WHEREAS OBJECTS STILL ON THE SURFACE OF A LARGE MASS IN A GRAVITATIONAL FIELD WILL.

6. OBJECTS ACCELERATING IN SPACE EXPERIENCE A FORCE WHEREAS OBJECTS ACCELERATING DOWN IN FREEFALL IN A GRAVITATIONAL FIELD DO NOT.

GENERAL RELATIVITY: STATIC INERTIAL REFERENCE BACKGROUND - "CURVED OR WARPED SPACE-TIME".

REALITY PHYSICS: AN ACTIVE INERTIAL REFERENCE BACKGROUND-IMPLOSIVE MOTION OF THE INERTIAL REFERENCE BACKGROUND DOWN THROUGH SPACE.

In Fig. 1 we now list six clues we use in Reality Physics to reveal the true cause of gravity. As shown in the top yellow box labeled number 1, we know gravity is a constant downward acceleration. In the yellow box 2 we have the second clue revealing that time slows down, or "dilates", in a gravitational field. In the yellow box 3 we have the third clue which says that all solid objects, no matter how heavy (neglecting any wind resistance), will always fall at the same rate. In the bottom yellow box labeled 4 we have the discovery that light bends around large objects, like planets or stars.

It is interesting to note here that, even though these four very informative clues have been around for more than a century, no one in Theoretical Physics Academia, or anywhere else, has any idea as to how to use them to discover the true cause of gravity. However, once we make the universe "active", as opposed to the "static" universe as miss assumed today, we enter a completely new realm of possibilities not even imagined by Theoretical Physics Academia

We now combine these four observations describing the true nature of gravity with these two other observations here, as taken from Einstein's Principle of Equivalence, as shown in the two orange boxes: 5 and 6. In the top orange box number 5, we reveal how an object stationary in space, located a great distance away from a gravitational field, will <u>NOT</u> experience any force, where an object located still of the surface of a large massive object <u>WILL</u>. In the orange box number 6 we now reveal the second part of Einstein's Principle of Equivalence which says that all objects accelerating down through space <u>DO</u> experience a force, whereas all objects accelerating downward through space in freefall in a gravitational field <u>DO NOT</u>.

By understanding why this is true we can begin to get a real insight into what actually causes gravity. The "SECRET", if you will, is to realize that stationary three-dimensional space <u>DOES NOT</u> define the inertial reference background of the universe. This means that the still position in the inertial reference background of the universe can "**MOVE**," or accelerate down with respect to the still position of stationary space. This is, of course, not possible with a static inertial reference background as described by General Relativity.

In the blue box directly below the orange box numbered 6 we reveal the fact that General Relativity Theory defines the universe as being characteristic of a static inertial reference background where gravity must be defined as: "curved or warped space, or spacetime". In the blue box directly below it we show how Reality Physics describes the universe as being characteristic of an "<u>active</u>", as opposed to a "<u>static</u>", inertial reference background. This now means that the still position throughout the universe can no longer defined as simply existing linearly from one moment to the next in time, but is continually being formed at the Speed of Light. This allows us to define gravity as: "**the implosively accelerating motion of the still position of the inertial reference background down through space**". In order to understand what kind of an active universe would be necessary for this new definition of gravity to exist, we refer to the Light Cone Diagram of Relativity, and the Time Cone Diagram of Reality Physics illustrated in Fig. 2.

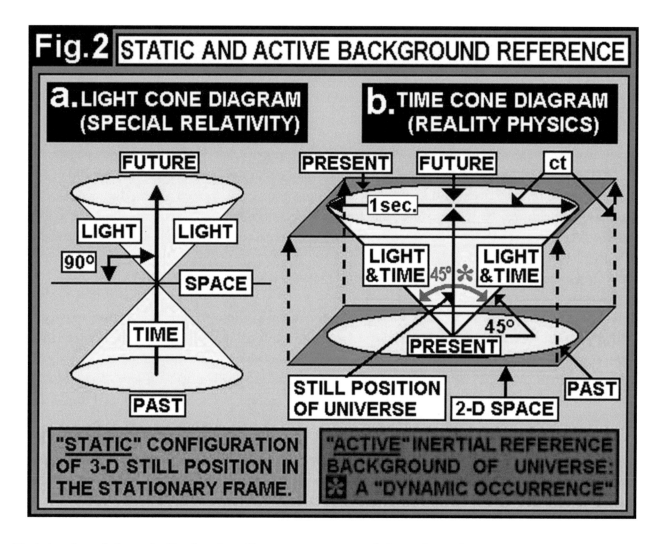

Fig.2 STATIC AND ACTIVE BACKGROUND REFERENCE

a. LIGHT CONE DIAGRAM (SPECIAL RELATIVITY)

FUTURE

LIGHT LIGHT

90°

SPACE

TIME

PAST

"STATIC" CONFIGURATION OF 3-D STILL POSITION IN THE STATIONARY FRAME.

b. TIME CONE DIAGRAM (REALITY PHYSICS)

PRESENT FUTURE ct

1 sec.

LIGHT & TIME 45° LIGHT & TIME

45°

PRESENT

STILL POSITION OF UNIVERSE

2-D SPACE PAST

"ACTIVE" INERTIAL REFERENCE BACKGROUND OF UNIVERSE: A "DYNAMIC OCCURRENCE"

To Aristotle and the early Greeks, the still position, as arranged throughout the entire universe, was physically defined as being a "static, three-dimensional, Euclidean type of configuration" where time simply passed linearly from one moment to the next. Sir Isaac Newton quickly adopted this static configuration and used it as the foundation concept upon which he based his ideas of absolute space and time, and his Three Laws of Motion.

Einstein, even though he did do away with both absolute space and absolute time with Relativity, still described the still position in the observer's stationary frame, from which all other relative motion is measured, as a static

6

configuration simply existing linearly from one moment to the next in time. We know this to be true because, as revealed in the left portion of Fig. 2a, the Time Vector in the Light Cone Diagram is at a 90-degree angle with respect to the Space Vector representing the still position of the observer and, according to Relativity, there are as many still frames of reference as there are stationary observers to view the relative motion of all other frames.

As shown by the small yellow box in the lower left-hand corner of the diagram, this static configuration of space and time is taken directly from the Galilean system of Classical Dynamics However, instead of using absolute space to define the still position with respect to which all motion is measured, Relativity uses the Speed of Light to construct an absolute stationary configuration of still position within observer's still frame with respect to which all other relative motion is measured. What this means is that space, time, mass and velocity must now all be defined as only relative concepts.

What we can now realize here is that it is this 2,500-year-old miss assumption, incorrectly describing the inertial reference background within the universe as a static condition simply existing linearly from one moment to the next in time., that is actually creating this intractable "ROADBLOCK" preventing all of Theoretical Physics Academia from ever being able to mathematically interconnect their version of Space-time Motion Physics (Special Relativity Theory) to Newtonian Dynamics. In other words, by miss assuming a static universe, they are simply placing themselves in the **wrong** universe.

This is easy to understand as we now take notice of the fact that the Time vector in the Light Cone Diagram is positioned in a vertical direction, meaning it is at a <u>right angle</u> to the vector representing space. This, as we observe, is taken directly from the diagram in the small yellow box describing the Galilean Transformations as shown in the small detail immediately to the left portion of the Light Cone Diagram describing the static universe of General Relativity.

So, how do we get into the right universe and expose the correct solution to our problem? We now totally reject this antiquated and incorrect version of the "static" universe as it is miss assumed to be today by the Light Cone Diagram of General Relativity. As we have revealed, this idea of a static universe, where time simply passes linearly from one moment to the next, goes back over 2,500 years as first visualized by Aristotle. This means that the still position of the universe remains the "<u>exact same place</u>" from one moment to the next. In our new universe of Reality Physics, as shown in Fig. 2b, the still position is continually being formed at the Speed of Light.

Fig. 2 (cont.) Referring back again to our light cone diagram of Relativity, we now move the Time vector from its original position pointing directly upward to a totally new configuration pointing outward at **45-degree angles** in both directions. This means that the Time vectors now exactly over-lay the paths of the Light vectors as shown by the Time Cone Diagram of Reality Physics, in the right-hand portion of Fig. 2b.

What we have just created here is a universe with an: <u>active,</u> as opposed to a: <u>static,</u> inertial reference background. This means that the still position within the universe does not just simply exist linearly from one moment to the next in time, as it is miss assumed to do today by Theoretical Physics Academia. It is an active entity continually being formed by the Speed of Time through Space, as Time now passes outward into Space, from every point in Space, with an active, two-dimensional, omnidirectional displacement at the Speed of Light, or: **"c" = 299,792,458 meters per second.**

This now illustrates how Reality Physics completely changes the universe from a static universe to an active one as we now reveal the "**Time Cone Diagram**". The important thing we notice here is that the Time vector, which was originally vertical, meaning that it is now at a 90-degree angle with respect to the horizontal plane representing Space, in the Light Cone Diagram of Relativity, has now <u>been moved outward 45 degrees in both directions</u>. This further means that these Time vectors now exactly overlay the 45- degree angled Light lines, further creating the condition where: <u>both Light and Time move out into space together at the Speed of Light, or "c"</u>. This now causes "still" position in space to continually move upwards in the Time Cone Diagram into the future at the Speed of Time, or Light, as indicated by the vertical dashed arrows extending upwards from the corners of the bottom plane to the top plane.

By letting the Speed of Time be equal to the Speed of Light we now create the condition where the still position of three-dimensional space now continually experiences an: "**active, two-dimensional, omnidirectional, omnipositional displacement of Time with Space at the Speed of Light throughout the entire universe**.

What this means here is that the three-dimensional still position, within which we all continually physically exist, and which extends throughout the entire universe is not "just here" from one moment to the next. It is <u>continually being formed, or REGENERATED at the Speed of Light by this outward motion of Time with Space.</u>

This now immediately explains to us why time continues to pass in the universe at the rate at which it does to physically turn a clock's hands. This also explains to us why, when we look outward into space we are also, simultaneously, looking back in time. This is because "**Now Points**" in time, with the information describing "events" originating at distant locations, simply travel from their point of origin to reach us as they move at the Speed of Light, or the Speed of Time, outward through space to our location here in space.

The best way to illustrate this is for us to imagine a room in three-dimensional space completely filled with almost an infinite number of microscopic light bulbs, each one being completely transparent to the light of all the others. If we switch on all of the light bulbs at the same instant, we would observe the light flash from each bulb to be a constantly increasing surface area of a sphere expanding outward at the Speed of Light into the room.

If we let the room represent space and the sphere of light emanating from each of the light bulbs represent Time, we have an analogy illustrating how Time passes outward into Space, from every point in Space, at the Speed of Light to continually form the inertial reference background, as it is described here by Reality Physics.

This now means we can no longer think of our universe as a static configuration simply existing linearly from one moment to the next in Time as it is currently miss understood by Theoretical Physics Academia. We must now completely redefine it as a: "**Dynamic Occurrence**", continually being formed at the Speed of Light, or Time, by this brand new active, two-dimensional, omnidirectional, omnipositional Speed of Time Through Space, as illustrated here in Fig. 2b. This new universe of active still position now completely replaces the static universe of Theoretical Physics Academia.

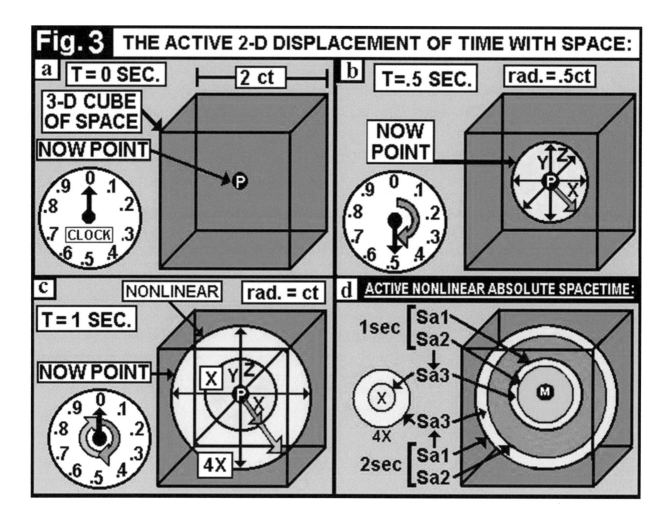

So now what we want to know is: How is it possible for this new description of the universe, as illustrated here by Reality Physics, to finally solve our problem of revealing the actual cause of gravity? To get the answer here we refer to Fig. 3 where we now illustrate how the actual, two-dimensional, outward, physical displacement, or motion, of Time passes outward into three-dimensional space.

In the diagram in the upper left-hand corner labeled "a" in Fig. 3 we now show a three-dimensional cube of space measuring two light seconds distance on either edge, with point P in the center. We let the "Now Point" in Time expand outward from center point P at the Speed of Light as indicated by the one second clock positioned in the lower left corner of the diagram as shown.

Just to the right of diagram 3a we now have diagram 3b which, as indicated by the small yellow sphere expanding omnidirectionally outward from point P, illustrates the active displacement of this Now Point in Time after **0.5 seconds** have passed, as also recorded on the clock in the left portion of 3a by the curved orange arrow.

Now, looking at diagram 3c in the lower left corner of Fig. 3 we observe that our clock now reads one full second as having passed. As indicated by the green arrow, this Now Point in Time has now expanded out to a sphere with a radius of twice that of the smaller yellow sphere, as shown by the larger concentric blue sphere that is surrounding it. As we can also notice here, the blue sphere for Time displacement is now four time larger than the smaller yellow sphere as the radius of the Time displacement doubles in size.

What this actually reveals to us here is that this continual outward motion of Time with Space is "**nonlinear**" in that is intersects more space per unit length of Time passing outward into space at the nonlinear rate at which the surface area of a sphere increases in size with respect to its radius, as this radius is increasing in size as it expands outward into Space at as the Speed of Light, or Time: "**c**".

In diagram 3d in the lower right corner of Fig. 3 we now show how a large body of mass "M", when located at point P, would cause a resistance to this outward motion of Time with Space. This now causes Time to <u>physically pass slower</u>, as indicated by the inner green circle (surface area two, or **Sa2**) representing the "Gravity Frame", than it does in "empty" Space, as indicated by the larger yellow circle around it (surface area one, or **Sa1**) representing the "Universe Frame".

Consequently, if we subtract **Sa2** from **Sa1** we obtain **Sa3**, which is the difference of these two surface areas after one second of Time has passed. In the left side of diagram 3d we show how, if we compare the size of **Sa3** at one second having passed, as shown by the small yellow circle, to **Sa3** after two seconds have passed, as shown by the larger concentric blue circle, we observe that the blue circle is <u>four times larger</u>.

We find that by analyzing this very simple observation we can now create an analogy illustrating how Reality Physics, because it uses an "<u>active, nonlinear time displacement with space</u>" rather than a "<u>linear time displacement</u>", is able to reveal the actual mechanism causing gravity, as illustrated next in Fig. 4 that follows.

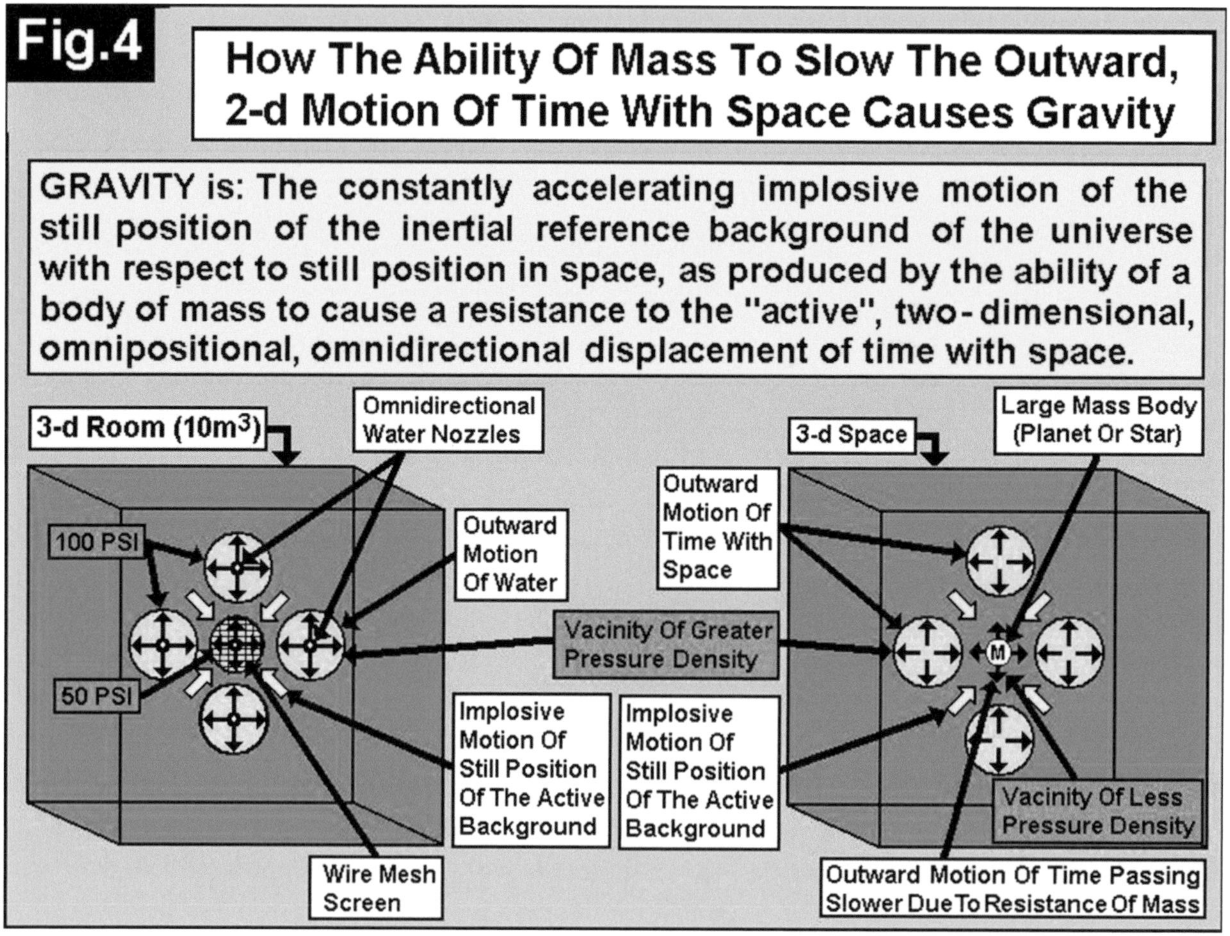

Fig.4 How The Ability Of Mass To Slow The Outward, 2-d Motion Of Time With Space Causes Gravity

GRAVITY is: The constantly accelerating implosive motion of the still position of the inertial reference background of the universe with respect to still position in space, as produced by the ability of a body of mass to cause a resistance to the "active", two-dimensional, omnipositional, omnidirectional displacement of time with space.

3-d Room (10m³)

Omnidirectional Water Nozzles

Outward Motion Of Water

100 PSI

50 PSI

Wire Mesh Screen

Vacinity Of Greater Pressure Density

Implosive Motion Of Still Position Of The Active Background

Implosive Motion Of Still Position Of The Active Background

3-d Space

Large Mass Body (Planet Or Star)

Outward Motion Of Time With Space

Vacinity Of Less Pressure Density

Outward Motion Of Time Passing Slower Due To Resistance Of Mass

For more than a century Theoretical Physics Academia has tried to use Einstein's General Theory of Relativity to explain gravity as "curved or warped space, or curved spacetime, etc." This attempt to explain gravity has resulted in absolutely no progress at all in revealing the actual cause of gravity. Why? Because Theoretical Physics Academia is in the wrong universe. This is due to the fact, as was revealed by the Light Cone Diagram as illustrated in the left portion of Fig. 2a, they incorrectly assume that the still position of three-dimensional space defines the inertial reference background of the universe as a static condition simply existing linearly from one moment to the next in time. So, they continually try unsuccessfully to explain gravity as the curving or warping of spacetime.

This is why we still have no idea at all as to what gravity actually is, or what causes it. What this means is that we have to completely discard General Relativity if we really want to know the truth about what gravity actually is, or its true cause. Gravity is an <u>action</u>, and therefore <u>must result from an action</u>, not just a curvature or warping of spacetime, as miss assumed by Einstein's General Relativity.

This now reveals to us why we must completely discard the Light Cone Diagram of Relativity, because it describes a static inertial reference background, as shown if Fig. 2a, and now replace it with the Time Cone Diagram of Reality Physics, as shown in Fig. 2b, if we ever want to know the real truth. We find that by using this new Time Cone Diagram of Reality Physics we go beyond the static universe of General Relativity. This now enables us to expose this new active universe, where the inertial reference background of the universe is not "just here" from one moment to the next for no apparent reason, but is continually being formed by the active, two-dimensional, outward displacement of Time with Space at "**c**".

In the large yellow box in the top portion of Fig. 4, we now reveal the true definition describing what gravity really is, as it is defined by this new universe of Reality Physics. Gravity is:

<u>The constantly accelerating implosive motion of the still position of the inertial reference background of the universe with respect to the still position of three-dimensional space as it is contained within the Absolute Universal Metric, or AUM, as produced by the ability of a body of mass to physically cause a resistance to this active, two-dimensional, omnidirectional, omnipositional displacement, or motion, of Time with Space at the Speed of Light, or Time.</u>

In the bottom of Fig. 4 we now have an analogy that enables us to understand how this ability of a body of mass to slow the outward, two-dimensional motion of time causes gravity. In the bottom left portion of Fig. 4 we have a blue cube representing a typical room, ten meters on each edge.

We now fill this room with omnidirectional hose nozzles one meter apart from which water flows outward in all directions at 100 P.S.I., and then exits the room from all sides. If we place a small object, say a ball with a specific gravity of one inside the room, it would bounce around the room, following what ever eddies and currents are produced. However, it would keep moving around in the room because the active pressure density is the same throughout the room.

Fig. 4 (cont.) We now place a wire mesh screen around the nozzle directly in the center of the room to effectively reduce the pressure of this middle nozzle from 100 P.S.I. to 50 P.S.I., as illustrated.

Obviously, this reduction of the water flow from this middle nozzle would cause a vicinity of less active pressure density in the middle of the room as compared to the greater pressure density throughout the rest of the room, continually being formed by the other nozzles in the room that are still at their greatest rate of 100 P.S.I.

This would also further cause an implosive motion of the water flow from the outer vicinities of the room, where the pressure is greater, toward the middle nozzle, where the pressure is less. This is indicated by the yellow arrows pointing inward from the light blue circles towards the middle nozzle covered with the wire mesh screen causing this difference in pressure density within the room.

From this example here can we observe how the ability of the wire mesh screen to slow the outward motion of the water flow would cause the implosive motion of the active pressure density of the water towards the nozzle located in the middle of the room. This would cause our object, with a specific gravity of one, to fall, or gravitate, towards the middle of the room where the pressure is less, as it moves away from the greater pressure in the rest of the room.

By our now letting the room represent three-dimensional space, the omnidirectional displacement of the water constantly expanding out into the room represent the outward motion of Time with Space, and the wire mesh screen represent the resistance of a body of mass to the outward motion of Time, we now create an analogy that will enable us the reveal the actual cause of gravity.

Now, looking at the blue cube in the bottom right portion of Fig. 4, which represents three-dimensional space, we observe that a body of mass, such as a planet or star, placed at the center of this cube of space at point "M" would cause a resistance to this outward motion of Time with Space in the same way the wire mesh screen causes a resistance to the outward flow of water into the room, as shown.

This resistance to the outward motion of time would obviously create a vicinity of "less active pressure density" at that location of the body of mass as compared to the greater active pressure density throughout the rest of space where time continues to pass at its greatest rate, as indicated by the large light blue circles.

Consequently, because we are dealing with an active, as opposed to a static, inertial reference background here, the effect of creating less active pressure density at the location of the body of mass, as compared to the greater active pressure density that is continually being created throughout the rest of space, would obviously cause the still position located within this inertial reference background to implode downward through space to the location of less active pressure density, where the body of mass is positioned.

This is, again, indicated by the small yellow arrows, pointing in a downward direction away from the greater pressure density existing throughout the rest of space, and down towards the location of the body of mass where the active background pressure density is less.

This means that small objects in "free fall" on the surface of a large body of mass <u>would remain at a still position within this active inertial reference background as they both fall down together, or gravitate, down towards the surface of the large body of mass.</u>

This now reveals to us here that gravity cannot be explained by the curvature or warping of space, or spacetime, as it is currently miss assumed to be by the static universe of General Relativity. It must now be completely redefined as:

The downwardly accelerating implosive motion of the inertial reference background of the universe with respect to the still position of the three-dimensional space within the Absolute Universal Metric as produced by the ability of mass to cause a resistance to the active, two-dimensional, omnidirectional, omnipositional displacement of Time with Space, as illustrated by the diagram located in the bottom right-hand portion of Fig. 4, as shown. As was stated before, gravity is an action, not a curvature or warping of something, and therefore must be caused by an action.

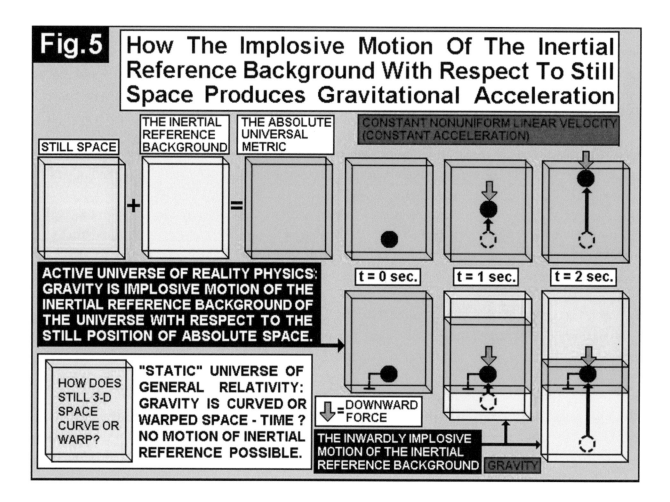

In Fig. 5 we now illustrate how this implosive motion of this still position within this inertial reference background with respect to the still position in three-dimensional space now produces gravitational acceleration as defined by this new concept of <u>ACTION GRAVITY</u> as revealed here by Reality Physics.

In the upper left portion of Fig. 5, we now show a blue rectangle representing still position in space. Directly to the right of it we have a yellow rectangle representing the stationary position within the universe's inertial reference background that is continually being formed by the outward motion of Time with Space.

The first thing we realize here is that accelerating motion of an object with respect to the still position, as it is defined by the inertial reference background (yellow rectangle), <u>does</u> cause a force to be produced, whereas accelerating motion with respect to the still position of space (blue rectangle), <u>does not</u>.

We will now three-dimensionally align, or superimpose, the still position, as it is defined by the inertial reference background of the universe, as represented by the yellow rectangle, within the blue rectangle representing space, to now result in the green rectangle, which represents the stationary position of the Absolute Universal Metric (or the AUM) positioned directly to the right of the yellow rectangle, as shown. In the green rectangle directly to the right of it we now place a small round object, as indicated by the black ball near the bottom where "t" equals zero seconds, as illustrated.

Now we let to small object move in an upward direction with a constant acceleration as illustrated by the vertical black arrow in the green rectangles representing the Absolute Universal Metric of the universe. This would obviously cause a force to be produced since the small round object is now <u>simultaneously</u> moving with respect to both, the stationary position of space, and the stationary position of the inertial reference background, which now exists as a plenum structure <u>exactly three-dimensionally aligned within space</u>.

So, since the small round object, by now moving through space, is also simultaneously moving with respect to the still position of the inertial reference background, it must then produce a force in the opposite direction of its acceleration, indicated by the small orange arrows pointing in the downward direction at: **t=1 sec**, and **t=2 sec.**

This immediately explains to us how the constant nonuniform linear velocity of an object in a remote location in space far away from any gravitational fields will produce a force. This is because, as we observe, it is not only accelerating with respect to the still position of space, but also with respect to the stationary position of the inertial reference background that is <u>three-dimensionally aligned with space</u>, as shown where "t" equals one and two seconds.

We find we can use this new observation to completely disprove Einstein's Principle of Equivalence and reveal this new: "Principle of Absolute Equivalence" of Reality Physics that, for the first time ever in the entire history of gravitational study, now enables us to mathematically interconnect gravitational force to centrifugal force.

Fig. 5 (cont.) As we can all observe by looking at the right portion of Fig. 5, we create this <u>same exact</u> <u>accelerating motion</u> by locating a body of mass at a still position within the space of the Absolute Universal Metric as shown in the green rectangle at "**t**" equals zero seconds. We then let the stationary position of the inertial reference background, indicated by the yellow rectangle, <u>implode down with an accelerating motion</u> past it at "**t**" equals one and two seconds.

This very simple demonstration here, for the first time in the history of all human scientific human investigation, now reveals to us the **actual mechanism** creating gravitational force. <u>The secret is for us to understand that</u> <u>the still position within the inertial reference background of the universe can move, or accelerate downward,</u> <u>with respect to the still position of three-dimensional space.</u>

What we have to realize here is that this condition where the still position within the inertial reference background can now move with respect to the still position of space <u>can only occur in an active universe as described by</u> <u>Reality Physics. It cannot possibly occur in the static universe as incorrectly projected by General Relativity</u> which, as shown in the lower left corner of Fig 5, miss assumes that still position in space defines the inertial reference background of the universe. This is because Relativity Theory uses the Light Cone Diagram, meaning that Physicists today know nothing about the active, two-dimensional, outward motion of Time with Space at "c", as depicted in the Time Cone Diagram of Reality Physics. So, as we can see, before you can have gravity, you have to have an outward motion of Time with Space to create an active, inertial reference background pressure density throughout the universe.

As previously illustrated in the lower right portion of Fig. 4, by placing a large body of mass, such as a planet or star, etc., into this active inertial reference background of Reality Physics, we now cause a resistance to this outward motion of Time with Space. This now further causes the still position, as located within this inertial reference background of the universe to accelerate in a downward direction with respect to the still position of three-dimensional space located at the large object's surface.

As illustrated here in the lower right-hand corner of Fig. 5, it is this constant, downwardly, accelerating motion of the inertial reference background, as indicated by the yellow rectangle, with respect to the still position in space, as indicated by the blue triangle, that now creates the gravitational acceleration.

Small objects located on the surface of a large body of mass would experience a downward force since the still position of the inertial reference background is now <u>accelerating downward past them</u>. These same small objects, if in free fall, would move along with the still position of the inertial reference background as shown by the clear yellow circles in Fig. 5.

As we will remember from clues #5 and #6 from Fig. 1, these two important observations about gravity have never been completely explained. How is it possible for an object to remain stationary with respect to the still space on the surface of a large body of mass and therefore experience a force, whereas an object stationary within the still position of "empty" three-dimensional space does not? General Relativity does indeed give us an observation, (Einstein's Principle of Equivalence), but it does not furnish us with an explanation.

Why? Because with this static universe, as it is incorrectly described by General Relativity, still space is incorrectly miss assumed to be the actual inertial reference background of the universe. What this further means is that most all of Theoretical Physics Academia today tries unsuccessfully to define gravity as the result of "curved or warped space", or "curved or warped spacetime".

What they don't realize here is that we all live in an "active", as opposed to a "static", universe where the still position in the inertial reference background of the universe can actually **physically move (accelerate down)** with respect to the still position of three-dimensional space of the Absolute Universal Metric.

This means that the real reason a force is produced for an object located at a still position on the surface of a large body of mass, even though it is positioned at a stationary location in still three-dimensional space, is because the inertial reference background is **accelerating down** past them, now creating the gravitational force.

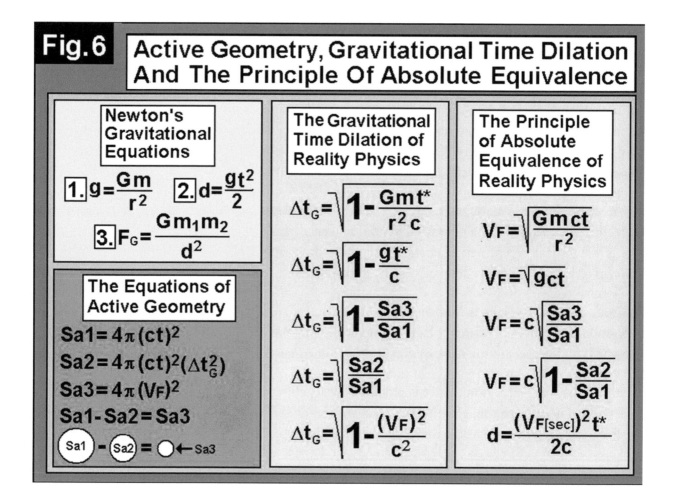

In the yellow box in the upper left-hand corner of Fig. 6 we now have Newton's three fundamental gravitational equations describing **1. Gravitational acceleration in terms of an object's mass and radius, 2. The distance an object will fall in time "t" due to gravitational acceleration,** and **3. The amount of gravitational force between two objects due to their mass and the distance between them.**

What we all need to understand here is that any correct description of the universe must always provide an internally consistent system of interconnected mathematical definitions that completely explains the origin and the application of these equations. The point that we really need to get across here is that any mathematical system that attempts to reveal the true nature of gravity must be one-hundred percent compatible with

Newtonian classical dynamics. This means that Einstein's Principle of Equivalence must now be extended to cover, not only constant nonuniform linear velocity, or constant acceleration, but also rotary motion and centrifugal force.

We now begin by introducing the Equations of Active Geometry of Reality Physics as shown in the orange box in the lower left-hand corner of Fig. 6. These four very simple equations mathematically describe the physical relationship of surface area one, or **Sa1**, which represents the rate at which time displaces space in the universe frame, to the smaller surface area two, or **Sa2**, which represents the displacement of time in the gravity frame, and to surface area three, or **Sa3**, which is the subtracted difference between **Sa1** and **Sa2**, as illustrated in the bottom portion of the orange box. These four new equations of Active Geometry, although they are indeed very simple are now what will make it possible for us, for the first time ever in the history of physics, to understand the actual, true cause of gravity

In the larger light blue box directly to the right we now show the equations of Gravitational Time Dilation of Reality physics. The first equation is in terms of mass and radius, and the one just below it is in terms of gravitational acceleration (where $t^* = 1$ sec. in both of these equations is necessary because we are no longer dealing with a "static" universe, but a new "active" one which uses Active Geometry, not current static geometry). The two equations directly below are in terms of Active Geometry, and the equation on the bottom is in terms of what is called the: **"Velocity Frame"** or **"VF"**

This Velocity Frame, as it is mathematically described here, equals the vector velocity that, when it is substituted into the Fitzgerald Formula as illustrated, results in a Time Dilation **exactly equal to** the Gravitational Time Dilation of the object we are working with. This, of course, now shows us how Gravitational Time Dilation is mathematically related to Time Dilation due to vector velocity.

In the large blue box in the right portion of Fig. 6 we show the equations of The Principle of Absolute Equivalence of Reality Physics. These five equations, as illustrated, now interconnect the Gravitational Time Dilation and the Active Geometry of Reality Physics to the Velocity frame now enabling us to reveal how this Principle of Absolute Equivalence allows us to expose how this ACTION GRAVITY is responsible for the origin of Newton's three gravitational equations, as shown in the small yellow box.

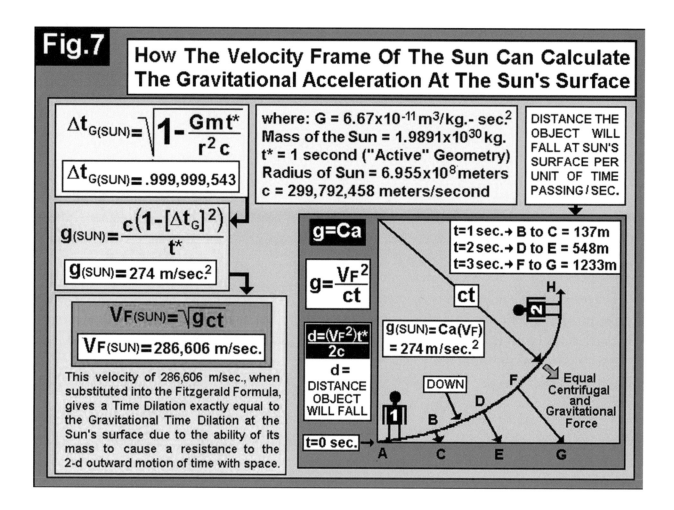

Fig.7 How The Velocity Frame Of The Sun Can Calculate The Gravitational Acceleration At The Sun's Surface

$$\Delta t_{G(SUN)} = \sqrt{1 - \frac{Gmt^*}{r^2 c}}$$

$$\Delta t_{G(SUN)} = .999,999,543$$

where: $G = 6.67 \times 10^{-11}$ m³/kg.- sec.²
Mass of the Sun = 1.9891×10^{30} kg.
t* = 1 second ("Active" Geometry)
Radius of Sun = 6.955×10^8 meters
c = 299,792,458 meters/second

DISTANCE THE OBJECT WILL FALL AT SUN'S SURFACE PER UNIT OF TIME PASSING / SEC.

$$g_{(SUN)} = \frac{c(1 - [\Delta t_G]^2)}{t^*}$$

$$g_{(SUN)} = 274 \text{ m/sec.}^2$$

$$V_{F(SUN)} = \sqrt{gct}$$

$$V_{F(SUN)} = 286,606 \text{ m/sec.}$$

This velocity of 286,606 m/sec., when substituted into the Fitzgerald Formula, gives a Time Dilation exactly equal to the Gravitational Time Dilation at the Sun's surface due to the ability of its mass to cause a resistance to the 2-d outward motion of time with space.

$$g = Ca$$

$$g = \frac{V_F^2}{ct}$$

$$d = \frac{(V_F^2)t^*}{2c}$$

d = DISTANCE OBJECT WILL FALL

t=0 sec.

$$g_{(SUN)} = Ca(V_F) = 274 \text{ m/sec.}^2$$

t=1 sec. → B to C = 137m
t=2 sec. → D to E = 548m
t=3 sec. → F to G = 1233m

DOWN

Equal Centrifugal and Gravitational Force

In Fig. 7 we now illustrate how this Velocity Frame (where VF of the Sun (**286,606 m/sec.**) is directly mathematically interconnected to the Gravitational Acceleration of the Sun. We begin in the small yellow box located in the upper left corner of Fig. 7 as shown. We now substitute in the mass and the radius of the Sun, as shown in the larger yellow box in the top middle portion of Fig. 7, into this equation to calculate the Gravitational Time Dilation of the Sun as shown in the small white box directly below the equation.

We now substitute this Time Dilation into the equation in the blue box that is located directly below to calculate the Gravitational Acceleration at the Sun's surface as shown in the smaller white box, which is: **274 m/sec.²**. When we then substitute this Gravitational Acceleration at the Sun's surface into the equation in the small orange box within the larger light blue box in the lower left corner of Fig. 7, we obtain the Velocity Frame corresponding to the Gravity Frame of the Sun, as shown in the small white horizontal box within the orange box to be **286,606 meters per second.**

The vector velocity of: "**VF$_{(SUN)}$**" equals **286,606 meters per second** and also equals that exact velocity which, when it is substituted into the Fitzgerald Formula, results in a Time Dilation that is exactly equal to the Gravitational Time Dilation of the Sun due to the ability of its mass to slow down time. This means that there are actually two different ways to slow down time in this universe. One way is to increase your vector, or orbital, velocity allowing you to "physically catch-up" to this active, two-dimensional motion of time that is passing omnidirectionally outward into space at "**c**". Or, we place a large mass body within the active pressure density as caused by the continual outward motion of time with space to create a vicinity of "less active pressure density" around the large body of mass, now resulting in what we call: "**Gravitational Acceleration**".

In the dark gray box in the lower right corner of Fig. 7 we show a diagram where we have two objects traveling at this velocity where: **VF$_{(SUN)}$ = 286,606 meter per second.** The first object travels along the straight path: **A-C-E-G**, and the second object takes the curved path: **A-B-D-F-H**, with a radius equal to "**ct**", as shown in the light gray square box. The object traveling along the straight line would obviously experience no force at all since it is not accelerating with respect to the still position of the inertial reference background of the universe, which is now three-dimensionally aligned with the still position of space that forms the Absolute Metric.

This would be analogous to an object in free fall at the Sun's surface since it too remains stationary with respect to a still position in the inertial reference background as it continually implodes with an accelerating downward motion where: "**g$_{(SUN)}$**" = **274 m/sec.^{2}** down towards the surface of the Sun.

As we also observe from this diagram, this means that this object in free fall at the Sun's Surface would fall a distance of **137 meters** at "**t**" equals one second, as shown at: "**B-C**", **548 Meters** at "**t**" equals two seconds, as shown at: "**D-E**", and a distance of: **1233 meters** at "**t**" equals three seconds, as shown at: "**F-G**", etc., as they appear in the small white box in the upper right-hand corner of the light gray box, as shown in Fig.7.

(**Fig. 7 (cont.)** As we observe from the gray box in the right-hand corner of Fig. 7, these distances of: **137 meters** at "**t**" equals one second from "**B** to **C**", **548 meters** at "**t**" equals two seconds from "**D** to **E**", and **1233 meters** at "**t**" equals three seconds from "**F** to **G**", are the **exact** distance the object turning in the circular path at this velocity frame of: "**VF = 286,606 meters per second**", will be displaced by the still position of the inertial reference background.

This now explains to us why an object turning in a circular path experiences this centrifugal force. It is because the "**ct**" vector is holding the object at a stationary position causing it to experience an inward acceleration away from the still position of the inertial reference background. This means that the still position within the inertial reference background will now accelerate outward from the moving object's still position causing an outward force (orange arrow) in the opposite direction of the inward acceleration of this object's rotary velocity, that is exactly equal to the Velocity Frame of the Sun which we found to be: **VF = 286,606 meters per second**.

This now shows us how this Principle of Absolute Equivalence of **ACTION GRAVITY** reveals to us how gravitational acceleration results directly from the implosive motion of the still position of the inertial reference background with respect to still position of space.

We can see how this would be true if we now visualize how a large body of mass holds a small object stationary on its surface the same as the "**ct**" **vector** holds our object in its circular path causing the still position located within the inertial reference background to continually displace it with an accelerating motion.

In the same way the surface of the Sun would hold a small object still in space at its surface as the stationary position in the inertial reference background accelerates with an implosive motion down past it. Subsequently, both the object turning in the circular path at this velocity of $\underline{VF_{(SUN)} = 286,606 \text{ meters per second}}$, and the object stationary of the surface of the Sun, would now both experience the still position of the inertial reference background to accelerate down past them at: **274 meters per second squared.**

This immediately explains to us why both of the small objects now experience a force since both are continually being displaced by the accelerating motion of the still position of the inertial reference background. This also explains to us why an object in free fall at the Sun's surface would fall the distance from "**B** to **C**" in one second, from "**D** to **E**" in two seconds, and "**F** to **G**" in three seconds, as shown in the diagram in the lower right corner of Fig. 7.

The observation we can obtain from the diagram in the lower right-hand corner of Fig. 7 allows us to understand how the force that is produced due to centrifugal acceleration originates from exactly the same cause as the force of gravity. The small object turning in a circular path is held in place by the "**ct**" vector meaning, that as it turns it must continually be physically displaced by the accelerating action of the still position of the inertial reference background, as shown. Accordingly, it is this still position, as it is located within this inertial reference background accelerating outward away from the object's still position, as the object turns in a circular path, that creates the centrifugal force experienced by the moving object.

Similarly, a small object located on the surface of a large body of mass would also experience a gravitational force since the still position of the inertial reference background accelerates down past it. This is because it is now continually being held at a stationary position in space by the surface of the large body of mass the same way the "ct" vector holds the small object stationary in its arc as it turns in its circular path. We now show how an observer would experience the same force whether or not the observer was standing on the surface of a planet (**1.**), or turning in a circular path (**2.**).

Both small objects, the one turning in a circular path as illustrated in Fig. 7, and the one stationary on the surface of the Sun, must now experience an equal force being produced since both are continually being displace with an accelerating motion of the still position of the inertial reference background, as it moves past them thereby creating the force they experience. What we have just illustrated here is <u>The Principle of Absolute Equivalence of Reality Physics</u>, as demonstrated by this new concept of: ACTION GRAVITY.

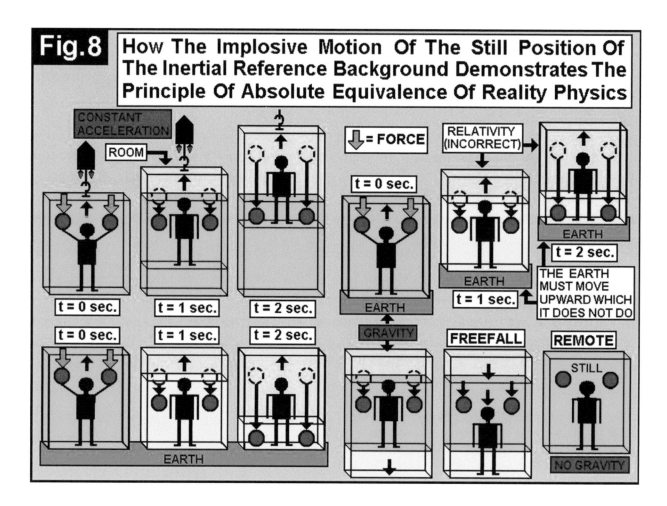

In Fig. 8 we now illustrate how we can use this new observation to prove that the Principle of Equivalence of General Relativity is completely invalid, while simultaneously revealing how this new Principle of Absolute Equivalence, as described here by the Action Gravity of Reality Physics, is completely correct. In the upper left-hand portion of Fig. 8, we now show an observer located stationary within this rectangularly shaped room that is three-dimensionally aligned with the still position of the absolute universal metric, as represented by the green rectangle at "**t**" equals zero seconds.

We now let a small rocket directly above the room accelerate the room in an upward direction, as indicated by the blue rectangle, as "t" equals one and two seconds as is illustrated. Obviously, if the observer released the two round objects they would immediately fall to the floor since they would always remain at a stationary position in the inertial reference background of the universe which is three-dimensionally aligned with the still position of three-dimensional space, as both the room and the observer, now experience a constant upward acceleration with respect to it.

Now, turning our attention to the upper right-hand corner of Fig. 8, we show how Einstein tries to produce this same exact situation in a gravitational field. According to Einstein's Principle of Equivalence of General Relativity Theory, the observer located stationary within a room on the surface of the Earth at "**t**" equals zero would, just like the observer being pulled upward by the rocket at "**t**" equals zero, experience a force pulling the two round objects down towards the floor of the room, as indicated by the two orange arrows pointing downward. As the observer on the Earth's surface releases the two round objects, they fall to the floor of the room with a constant downward acceleration as illustrated at position "t" at one and two seconds, just like the observer being pulled upward by the rocket.

The Relativists will try to make the argument here that these two situations are completely equivalent but, as we can observe, there is an obvious problem. In order for this to be true the surface of the Earth, and the still position of space, would have to physically move in an upward direction with respect to the still position within the inertial reference background (the yellow triangle), and, as we all obviously know from our own personal experience, this <u>does not</u> happen! It is this very problem that, not only completely invalidates the Principle of Equivalence, but it also completely disproves the validity of Einstein's General Theory of Relativity.

Directly below, in the far-left green rectangular cube in Fig. 8, we show how an observer located on Earth would experience a force being produced as the stationary position of the inertial reference background accelerates down past them (orange arrows). As the small objects are released, they would always a fall to the ground with the exact same downward motion as depicted in the example above where the rocket causes the upward acceleration of the room.

We can see how this would be true by now turning our attention to the upper right portion of Fig. 8, as we let the room positioned on the Earth accelerate upwards through space to create the exact same situation where it appears that the room is accelerating upward with respect to the still position of the inertial reference background.

Fig. 8 (cont.) From Einstein's Principle of Equivalence, we know that the reason as to why a force is being produced here is because both of these situations are supposed to be equivalent. What we also know is that a force can only be produced when an object has an accelerating displacement with respect to the still position, as it is defined by the inertial reference background of the universe.

However, as we can immediately observe, in the situation where the rocket is pulling the room upward the still position within the room is now accelerating upward with respect to the still position of space as it is now exactly three-dimensionally aligned with the inertial reference background of the universe. This now explains why the two round objects remain stationary with respect to the still position within the inertial reference background as the room is accelerated upward through space, making it appear that the objects fall in a downward direction in the room, just as they do with gravity.

This is equivalent to the floor of the room moving upward through space as the still position of the inertial reference background now remains at a stationary position with respect to this still position of space. However, as also shown in the example in the top right portion of Fig. 8, this would mean that the surface of the Earth <u>would have to physically move upward with respect to both the still position of the inertial reference background as it is aligned with the still position of space</u>. Otherwise, how could the two round objects remain at a stationary position with respect to the inertial reference background as shown and fall to the floor of the room?

Since we all know very well that that the room is <u>NOT</u> accelerating upwards through three-dimensional space, then the only explanation we have left: <u>is that the inertial reference background is accelerating downward with respect to the room as it remains at a stationary position in space</u>. As we can further observe, this is because this is the only way to satisfy the condition where a force is produced in both of these situations. So, according to Einstein's Principle of Equivalence, the observer in his elevator is accelerating upward to create the downward force. However, we know that the surface of the Earth does not do this and therefore, no force can be created.

Now, looking at the lower left portion of Fig. 8, we illustrate the difference between the Principle of Absolute Equivalence of Reality Physics, as compared Einstein's Principle of Equivalence of General Relativity. With the Principle of Absolute Equivalence, we can observe that the still position in the inertia reference background <u>can accelerate downward with respect to the still position of space.</u>

This means that the Earth can stay in one piece as the two round objects accelerate downward at "t" equals one and two seconds as shown. As also shown in the lower right-hand corner of Fig. 8, we illustrate how these two round objects in free fall would remain still with respect to the observer just as they would if located in a room in a remote vicinity of space located far away from any gravitational fields. From this demonstration here we observe why we must now totally discard Einstein's Principle of Equivalence, and replace it with this new Principle of Absolute Equivalence of Reality Physics.

As was previously revealed, the real reason that General Relativity cannot solve these contradictions by using Einstein's Principle of Equivalence, is because it describes a static universe where the still position within the inertial reference background <u>cannot move</u> with respect to the stationary position in space. However, with the active universe of Reality Physics, we now observe that, since the inertial reference background of the universe is an active entity, it <u>can</u> have an accelerating motion with respect to the still position of space.

This is what allows the Principle of Absolute Equivalence, as it is defined by the Active Geometry of Reality Physics, to solve this ridiculous contradiction existing within Einstein's Principle of Equivalence, where the floor of the room must now accelerate upward to produce the gravitational acceleration. This solution can only occur in the active universe of Reality Physics, meaning it cannot occur in Theoretical Physics Academia because they remain within the static universe as it is described by General Relativity. By understanding how this Principle of Absolute Equivalence of Action Gravity solves this problem as revealed here, we are finally able to understand how the still position of the inertial reference background can move with respect to still space, and cause gravity.

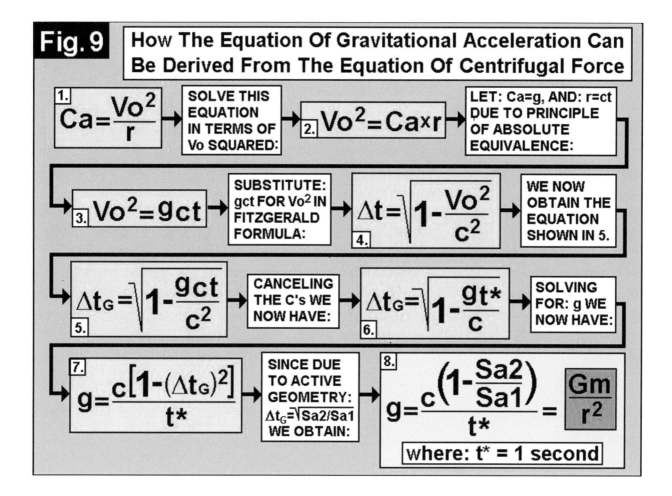

Fig. 9 — How The Equation Of Gravitational Acceleration Can Be Derived From The Equation Of Centrifugal Force

1. $$Ca = \frac{Vo^2}{r}$$

SOLVE THIS EQUATION IN TERMS OF Vo SQUARED:

2. $$Vo^2 = Ca \times r$$

LET: Ca=g, AND: r=ct DUE TO PRINCIPLE OF ABSOLUTE EQUIVALENCE:

3. $$Vo^2 = gct$$

SUBSTITUTE: gct FOR Vo^2 IN FITZGERALD FORMULA:

4. $$\Delta t = \sqrt{1 - \frac{Vo^2}{c^2}}$$

WE NOW OBTAIN THE EQUATION SHOWN IN 5.

5. $$\Delta t_G = \sqrt{1 - \frac{gct}{c^2}}$$

CANCELING THE C's WE NOW HAVE:

6. $$\Delta t_G = \sqrt{1 - \frac{gt*}{c}}$$

SOLVING FOR: g WE NOW HAVE:

7. $$g = \frac{c[1 - (\Delta t_G)^2]}{t*}$$

SINCE DUE TO ACTIVE GEOMETRY: $\Delta t_G = \sqrt{Sa2/Sa1}$ WE OBTAIN:

8. $$g = \frac{c\left(1 - \frac{Sa2}{Sa1}\right)}{t*} = \frac{Gm}{r^2}$$

where: t* = 1 second

In Fig. 9, we illustrate how this Principle of Absolute Equivalence allows us to derive this equation of Active Geometry of Reality Physics directly from Newton's equation mathematically defining centrifugal acceleration, now enabling us to correctly calculate the actual gravitational acceleration at the surface of a body of mass.

In the upper left corner of Fig. 9 in the blue box labeled # 1, we show how we solve Newton's Equation for centrifugal acceleration in terms of velocity squared, as shown in box # 2. We now, as allowed by the Principle of <u>Absolute</u> Equivalence, let "r", or radius, equal "**ct**", and "**Ca**", or centrifugal acceleration, be equal to "**g**", or gravitational acceleration, to obtain the equation shown in box # 3.

We now substitute this "**gct**" term into the Fitzgerald Formula for "**Vo**", as shown in box # 4 to now result in the equation as shown in box # 5. Canceling the "**c's**", as shown in box # 6, and then solving this equation for "**g**", we obtain the equation in box # 7, as shown.

Now, we must go all the way back to Fig. 6 as we turn our attention to the equation as shown in the second from the bottom position of the large blue box in the middle of Fig. 6 where we now have the equation: $\Delta t_G = (Sa2/Sa1)^{1/2}$. This is an equation of Active Geometry that mathematically defines Gravitational Time Dilation in terms of: **Δt sub-G**, which is equal to the square root of surface area two divided by surface area one. We now substitute this square root of **Sa2** divided by **Sa1** for the: **Δt sub-G** term in equation # 7 in Fig. 9 to now end up with the equation of Active Geometry as revealed in the large yellow box # 8 located in the bottom right corner of Fig. 9.

This equation now enables us to correctly calculate the gravitational acceleration at the surface of a body of mass in terms of the ratio of surface area one (with a radius of: "ct", which now represents the "Universe Frame"), and surface area two (with a radius equal to the distance the "Now Point" in Time will expand outward into space in the "Gravity Frame"). This equation now allows us to explain the origin of Newton's Equation mathematically defining gravitational acceleration in terms of mass and radius, as illustrated in the small orange box inside of the larger yellow box in the bottom of Fig. 9.

The important thing we need to understand here is <u>that the ratio of these two surface areas must always remain the same meaning that the gravitational acceleration that results must always be "constant".</u> This is because each of these surface areas represents a non-similar inertial frame of reference. Since we all know that the difference between any two non-similar inertial frames is an acceleration, and we also know that this difference must always remain a constant ratio as these two surface areas of time displacement continually expand outward, we now understand why Newton's Equation in the small orange box within the larger yellow one describes gravity as a <u>constant</u> acceleration. With this very important observation in mind, we begin to realize that it is this ability of a body of mass to slow down this active, two-dimensional, omnipositional, omnidirectional, motion of Time with Space that actually causes gravity, as we can now mathematically prove with the equation, we have in Fig. 10 that immediately follows.

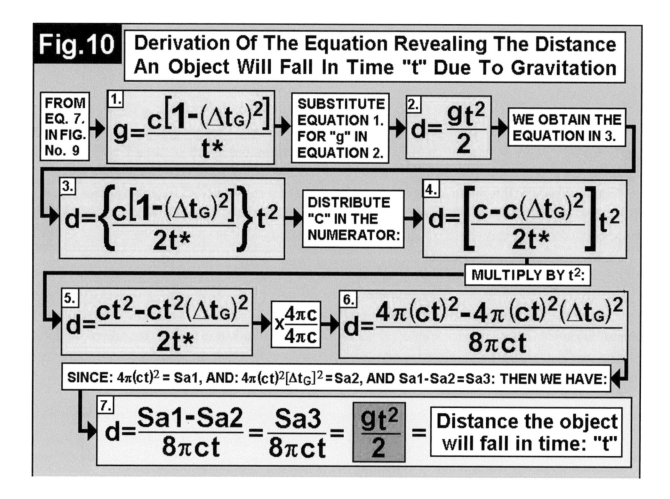

Fig. 10 — Derivation Of The Equation Revealing The Distance An Object Will Fall In Time "t" Due To Gravitation

In Fig. 10, we now illustrate how we can derive the equation of Active Geometry, as mathematically described by this new Action Gravity of Reality Physics, that enables us to correctly calculate the distance an object will fall in a given unit length of time: "**t**" in a gravitational field, as shown by the large yellow box in Fig. 10.

We start by looking at the blue box labeled # 1 in the upper left corner of Fig. 10, as we now take the term from the right-hand side of the equation as shown in box # 7 in Fig. 9, and substitute it for the "**g**" term in box # 2. This now gives us the equation as shown in box # 3. We now distribute the "**c**" term to result in the equation in box # 4. Now, multiplying the term on the right side of this equation by "**t²**", we arrive at the equation in

32

box # 5. We now multiply the right side of this equation by <u>one</u>, which we can write as: **4πc / 4πc**, as shown, to now obtain the equation in box # 6.

As was revealed in the orange box in the lower-left corner of Fig. 6, **Sa1** (Universe Frame) equals **4π(ct)²**, and **Sa2** (Gravity Frame) is equal to **4π(ct)² (Δt$_G$)²**, or the Gravitational Time Dilation squared, we end up with the equation in the left-hand portion of the large yellow box at the very bottom of Fig. 10 where we now observe that **d = distance object will fall = (Sa1-Sa2)/8πct, or Sa3/8πct.**

Since we know from the orange box back in Fig. # 6 that surface area two subtracted from surface area one equals surface area three, we find we are now able to obtain this very simple equation just to the right which divides surface area three by **8πct** to now correctly calculate the distance an object will fall per unit length of time "**t**" in a gravitational field. As we can notice here, this **8πct** term is the derivative of the right side of the equation where surface area=**4πr²**.

This equation allows us to explain the origin of Newton's Equation in the small orange box mathematically defining the distance an object will fall per unit length of time in terms of Gravitational Acceleration. This is because, as the radii of these two surface areas (**Sa1 and Sa2**) increase in linear increments, their difference in surface area increases as the square of the radius.

This means that as we change the "**t²**" term in Newton's Equation from one to two to three seconds, and even higher (even though the rate at which these two surface areas of Time Displacement expands always equals the exact same ratio), the <u>difference</u> between them will always continually increase by the nonlinear rate from four to nine to sixteen, and so on as time continues to pass, because it is a <u>difference</u>, and not a <u>quotient</u>, we are dealing with here.

So, as we observe, it is the difference in these two surface areas as indicated by surface area three that determines the actual distance an object will fall per unit length of time passing in a gravitational field. Since we know that the rate at which these spheres of time displacement diverge is directly dependent upon the amount of resistance the large mass body has to the outward motion of time with space, we observe how this simple equation proves this new Action Gravity as mathematically described here by Reality Physics

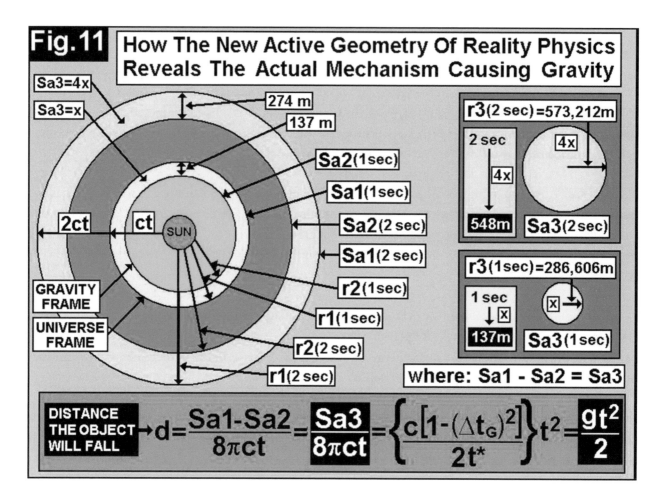

Fig. 11 — How The New Active Geometry Of Reality Physics Reveals The Actual Mechanism Causing Gravity

Sa3=4x
Sa3=x
274 m
137 m
Sa2(1sec)
Sa1(1sec)
Sa2(2 sec)
Sa1(2 sec)
r2(1sec)
r1(1sec)
r2(2 sec)
r1(2 sec)
2ct
ct
SUN
GRAVITY FRAME
UNIVERSE FRAME

r3(2 sec)=573,212m
2 sec
4x
4x
548m
Sa3(2 sec)

r3(1sec)=286,606m
1 sec
x
137m
x
Sa3(1sec)

where: Sa1 - Sa2 = Sa3

DISTANCE THE OBJECT WILL FALL →

$$d = \frac{Sa1-Sa2}{8\pi ct} = \frac{Sa3}{8\pi ct} = \left\{\frac{c[1-(\Delta t_G)^2]}{2t^*}\right\}t^2 = \frac{gt^2}{2}$$

In the left-hand portion of Fig. 11 we now introduce a new type of diagram that is only in Reality Physics called an: "Active Geometry Diagram". We find we can use these diagrams to illustrate how this new Active Geometry of Reality Physics now reveals the actual mechanism causing gravity. As shown by the small orange circle in the center of the diagram, we use the Sun as our example.

We now go back to the equation as illustrated in the very top of the left blue box in Fig. 6 mathematically defining the Gravitational Time Dilation in terms of both mass and radius. Now, substituting the mass of the Sun which is: **1.9891x10³⁰ kg.**, and the radius of the Sun which is: **6.995x10⁸ meters**, into this equation we find we are now able to obtain the correct Gravitational Time Dilation of the Sun to be: $\Delta t_{G(SUN)} = .999,999,543$.

Multiplying this Gravitational Time Dilation times, the linear distance Time or Light, will travel in one second, or "**ct**", we obtain the radius for $Sa2_{(SUN)}$ for one second of time passing to be: **radius of $Sa2_{(1\ sec.)}$ = 299,792,321 meters**.

We square this radius of: **299,792,321 meters**, and multiply it by "**4π**" to calculate the surface area of $Sa2_{(SUN)}$, after one second of time has passed to be: **$1.129408033479 \times 10^{18}$ meters^2**, as indicated by the green circle in the diagram representing the "Gravity Frame of the Sun". We now subtract $Sa2_{(SUN)}$, from **Sa1**, as represented by the larger yellow circle in the Active Geometry Diagram to obtain: **$Sa1 = 1.129409066758 \times 10^{18}$ meters^2**, to find their difference of **Sa3**.

This now allows us to calculate the difference between **Sa1** and **Sa2** resulting in a $Sa3_{(SUN)}$ at one second to be **$1.03227965 \times 10^{12}$ meters^2** as shown in the yellow area in the diagram, and also by the small yellow circle in the detail in the upper right portion of Fig. 11. As shown by the equation in the orange box in the middle of the large gray horizontal box in Fig. 11, we can divide this $Sa3_{(SUN)}$ at one second by "**8πct**" to obtain the **137 meter** distance the object will fall at the Sun's surface after one second of time has passed to equal: [**d: distance the object will fall = $Sa3_{(SUN)}/8πct = 137$ meters**].

Looking back to our Active Geometry Diagram in the upper left portion of Fig. 11, we observe that as two seconds pass these two surface areas of time displacement: $Sa2_{(SUN)}$ and **Sa1** will expand out to twice the radius they were at "**t**" equals one second. This means their surface areas increase in size by a factor of four as shown by the blue concentric circle on the outside of the diagram.

This also means that the subtracted difference between both of these two surface areas, or $Sa3_{(SUN)}$, will become four times larger at two seconds than it was at one second or **4.1291168×10^{12} meters^2**, as indicated by the light blue circle on the outside of the diagram. This is also indicated by the light blue circle in the small detail in the lower right-hand portion of Fig. 11, showing that the small object would fall <u>four</u> times further in two seconds, or **548 meters**, and <u>nine</u> times further in three seconds, or **1,233 meters**, than it did at one second. This now means that: "**<u>the rate at which an object falls exactly follows the rate at which Sa3 increases in size, now explaining how this ACTION GRAVITY uses this new Active Geometry of Reality Physics to reveal the true cause of gravity for the first time ever in the entire history of physical science.</u>**

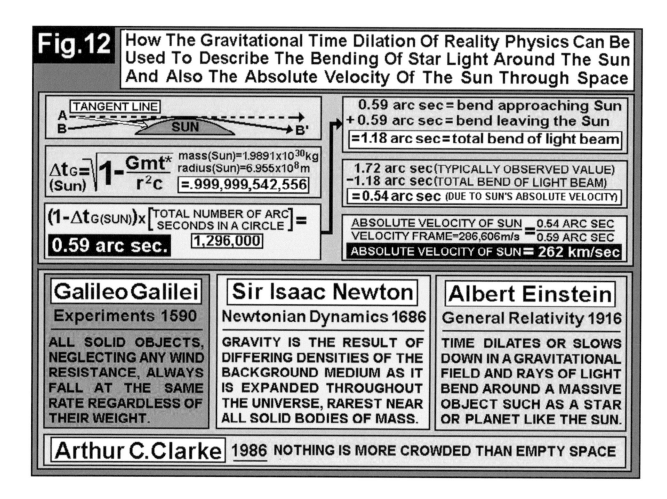

Fig. 12 — How The Gravitational Time Dilation Of Reality Physics Can Be Used To Describe The Bending Of Star Light Around The Sun And Also The Absolute Velocity Of The Sun Through Space

TANGENT LINE

A
B
SUN
B'

$$\Delta t_{G\,(Sun)} = \sqrt{1 - \frac{Gmt^*}{r^2 c}}$$

mass(Sun)=1.9891x10^{30}kg
radius(Sun)=6.955x10^8 m
=.999,999,542,556

$$(1 - \Delta t_{G(SUN)}) \times \left[\begin{array}{c}\text{TOTAL NUMBER OF ARC}\\ \text{SECONDS IN A CIRCLE}\end{array}\right] =$$

0.59 arc sec.   1,296,000

0.59 arc sec = bend approaching Sun
+ 0.59 arc sec = bend leaving the Sun
=1.18 arc sec = total bend of light beam

1.72 arc sec (TYPICALLY OBSERVED VALUE)
−1.18 arc sec (TOTAL BEND OF LIGHT BEAM)
= 0.54 arc sec (DUE TO SUN'S ABSOLUTE VELOCITY)

$$\frac{\text{ABSOLUTE VELOCITY OF SUN}}{\text{VELOCITY FRAME}=286,606 m/s} = \frac{0.54\ \text{ARC SEC}}{0.59\ \text{ARC SEC}}$$

ABSOLUTE VELOCITY OF SUN = **262 km/sec**

Galileo Galilei	Sir Isaac Newton	Albert Einstein
Experiments 1590	Newtonian Dynamics 1686	General Relativity 1916
ALL SOLID OBJECTS, NEGLECTING ANY WIND RESISTANCE, ALWAYS FALL AT THE SAME RATE REGARDLESS OF THEIR WEIGHT.	GRAVITY IS THE RESULT OF DIFFERING DENSITIES OF THE BACKGROUND MEDIUM AS IT IS EXPANDED THROUGHOUT THE UNIVERSE, RAREST NEAR ALL SOLID BODIES OF MASS.	TIME DILATES OR SLOWS DOWN IN A GRAVITATIONAL FIELD AND RAYS OF LIGHT BEND AROUND A MASSIVE OBJECT SUCH AS A STAR OR PLANET LIKE THE SUN.

Arthur C. Clarke | 1986 NOTHING IS MORE CROWDED THAN EMPTY SPACE

In Fig. 12 we now illustrate how the Gravitational Time Dilation of the ACTION GRAVITY of Reality Physics can correctly describe the bending of a beam of starlight around the Sun, and also reveal: the absolute velocity of the Sun through space. In the blue box in the upper left corner of Fig. 12, we show a horizontal line "**A**" illustrating how a beam of starlight traveling in a tangent line barely grazing the Sun's surface would be pulled downward by the Sun's gravity directly into the Sun. This is indicated by the white arrow originating at point "A", as it falls down onto the Sun's surface

However, a beam of light originating from a point slightly below this tangent line, as shown at "B", would bend around the Sun to follow the path to "B prime", as shown. What we must keep in mind here is that this same amount of bend of the light beam must occur equally on both sides of the Sun to obtain the total amount of arc seconds the light beam must bend. This means we must double the amount we calculate to get the correct amount the beam will bend.

We find we can calculate this amount of arc second bend due to the Sun's gravity by, first of all, now using this new equation of Action Gravity mathematically defining Time Dilation in terms of mass and radius as shown in the horizontal blue box in the left portion of Fig. 12 directly below the top blue box. By substituting both the mass and the radius of the Sun into this equation we obtain a Time Dilation due to gravity for an object the exact size of the Sun to be: $\underline{\Delta t_{G(SUN)}} = \underline{.999,999,542,556}$, as shown in the smaller white box.

Subtracting this Gravitational Time Dilation of the Sun from one, as shown in the yellow horizontal box directly below it, and then by multiplying this result by the total number of arc seconds within a circle, which is: 1,296,000, we now end up with an answer here of: **.59 arc seconds**, as shown in the black box in the lower left corner of the yellow box, located in the middle left portion of Fig. 12.

Now, following the black arrow from the right side of the yellow box upward to the left side of the blue box in the upper right portion of Fig. 12, we observe how we must now add these two amounts of arc second bend together to obtain a total bend of the light beam grazing the Sun's surface to be: **1.18 arc seconds**. Subtracting this 1.18 arc second bend from the typically observer value of 1.72 arc seconds, we now obtain: **.54 arc seconds**, as shown in the small white box directly below.

We now, as shown in the yellow box directly below the blue box, make a very simple equation where we multiply the Velocity Frame of the Sun by this .54 arc seconds, then divide by: .59 arc seconds, to now obtain the Absolute Velocity of the Sun through space to be: **262 kilometers per second**.

There have been several attempts to measure the Absolute Velocity of the Sun (and the Solar System) through space and some of them have come very close to the velocity we have just calculated here. It will be interesting to see how future measurements will agree, or deviate, from this value we have arrived at here using this new Active Geometry of Reality Physics.

Fig. 12 (cont.) In the very bottom portion of Fig. 12 we have three boxes, each one describing the vital contribution to the discovery of the cause of gravity as revealed here by Albert Einstein, Sir Isaac Newton, Galileo, and the Futurist: Arthur C. Clarke.

As shown in the light blue box it was Albert Einstein who first discovered that time dilates, or slows down in a gravitational field, and that light bends around large massive objects like planets or stars. By now combining these observations of Einstein with the discovery of Reality Physics, where time expands outward into space, from every point in space, at the speed of light, we can now understand how a body of mass would cause a resistance to this outward motion of time. This would further cause time to physically pass slower in the vicinity of the body of mass as compared to the rate at which time is passing out into the empty space surrounding it, immediately explaining the real reason as to why time slows down in a gravitational field, as discovered by Einstein.

So, as we also observe here, according to the concept of Action Gravity, this slowing of time by mass would cause a vicinity of less active pressure density at that location of the body of mass as compared to the grater pressure density existing throughout the rest of the universe where time continues to pass at its fastest Rate. This means that we are now using the **resistance** of a body of mass to <u>slow-down</u> the outward two-dimensional motion of time with space to now create the still position within the vicinity of the large mass at a slower rate, and therefore at a "**lower pressure density**", than the surrounding still position throughout the rest of the universe.

This would now obviously cause the surrounding vicinities that are characteristic of a greater pressure density, to implode with an accelerating motion down towards the vicinity of less active pressure density where the large mass is located. This implosively accelerating motion of the inertial reference background down through space we call: "**Gravity**". It is interesting that this basic nature of gravity was actually discovered by Sir Isaac Newton over three and one-half centuries ago and here we all are, now finally "catching-up" to Newton, the greatest genius who ever lived!

Now, turning our attention to the yellow box directly to the left we immediately understand how this new observation, which describes gravity as differing densities of the background medium as revealed by Sir Isaac Newton more than three centuries ago, exactly agrees with this new discovery of Reality Physics showing how this is only possible with the active universe of Reality physics. This is because, unlike with the static universe of Theoretical Physics Academia, with an active background medium we find we can now use mass to slow-down the outward motion of time with space. Newton, like everybody either before, or after, also miss assumed the universe to be a static condition simply existing from one moment to the next in time and consequently was never able to understand how this brand-new universe of Reality Physics actually created these: "<u>differing densities in the inertial reference background</u>" which are responsible for producing the **Gravitational Acceleration**.

This now brings us to the orange box in the lower left-hand corner of Fig. 12 where we have Galileo's discovery, made more than four centuries ago, revealing that <u>all objects, neglecting air resistance, fall at the same rate</u>. This is explained by the discovery of Action Gravity showing how the stationary position of the inertial reference background implodes downward with a <u>concentrically implosive accelerating motion</u> towards the surface of a large body of mass. All objects at the surface of a large body of mass would fall at the exact same rate since they are "<u>all moving along with the concentrically imploding sphere of the stationary position in the inertial reference background as it falls down through space towards the surface of the large body of mass</u>, as revealed here by Action Gravity.

At the very bottom of Fig. 12, we now have the "Futurist": Arthur C. Clarke, along with his very formidable quote: "<u>Nothing is more crowded than empty space</u>". Although this might sound completely counter intuitive to most all of us, what we must understand here is that: **he was exactly right!** The outward motion of time occurs at its greatest rate in what we perceive as "<u>empty space</u>". If we locate a large body of mass within this "active" outward motion of time with space it slows down, and this is what causes: **Gravity**.

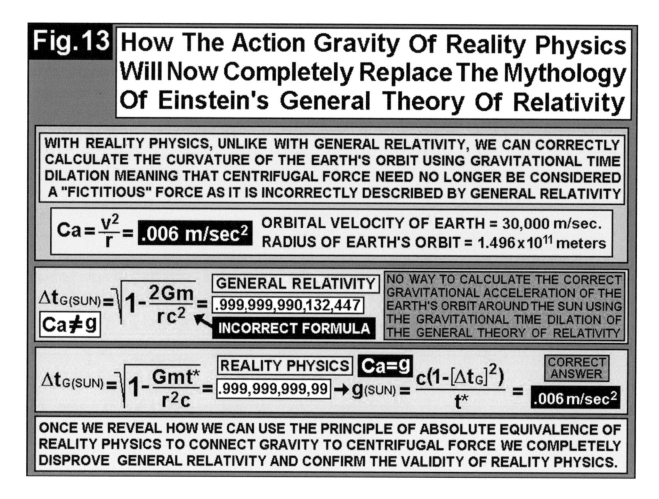

Fig. 13 How The Action Gravity Of Reality Physics Will Now Completely Replace The Mythology Of Einstein's General Theory Of Relativity

WITH REALITY PHYSICS, UNLIKE WITH GENERAL RELATIVITY, WE CAN CORRECTLY CALCULATE THE CURVATURE OF THE EARTH'S ORBIT USING GRAVITATIONAL TIME DILATION MEANING THAT CENTRIFUGAL FORCE NEED NO LONGER BE CONSIDERED A "FICTITIOUS" FORCE AS IT IS INCORRECTLY DESCRIBED BY GENERAL RELATIVITY

$$Ca = \frac{v^2}{r} = \boxed{.006 \text{ m/sec}^2}$$ ORBITAL VELOCITY OF EARTH = 30,000 m/sec. RADIUS OF EARTH'S ORBIT = 1.496×10^{11} meters

$$\Delta t_{G(SUN)} = \sqrt{1 - \frac{2Gm}{rc^2}}$$ $Ca \neq g$ GENERAL RELATIVITY $= \boxed{.999,999,990,132,447}$ INCORRECT FORMULA

NO WAY TO CALCULATE THE CORRECT GRAVITATIONAL ACCELERATION OF THE EARTH'S ORBIT AROUND THE SUN USING THE GRAVITATIONAL TIME DILATION OF THE GENERAL THEORY OF RELATIVITY

$$\Delta t_{G(SUN)} = \sqrt{1 - \frac{Gmt^*}{r^2c}}$$ REALITY PHYSICS $\boxed{= .999,999,999,99} \rightarrow$ $Ca = g$ $g_{(SUN)} = \frac{c(1-[\Delta t_G]^2)}{t^*} = \boxed{.006 \text{ m/sec}^2}$ CORRECT ANSWER

ONCE WE REVEAL HOW WE CAN USE THE PRINCIPLE OF ABSOLUTE EQUIVALENCE OF REALITY PHYSICS TO CONNECT GRAVITY TO CENTRIFUGAL FORCE WE COMPLETELY DISPROVE GENERAL RELATIVITY AND CONFIRM THE VALIDITY OF REALITY PHYSICS.

In Fig. 13 we reveal direct absolute proof of how this new concept of Action Gravity of Reality Physics allows us to totally disprove Einstein's General Theory of Relativity. We accomplish this by illustrating how we can correctly calculate the known curvature of the Earth's orbit by using the Gravitational Time Dilation of the new concept of Action Gravity. We prove this by demonstrating how the known Gravitational Acceleration of the Sun, as it is calculated by the Active Geometry of the Principle of Absolute Equivalence of Reality Physics, is <u>exactly equal</u> the centrifugal acceleration of the Earth, as it is also calculated by Newton's equation mathematically describing the centrifugal acceleration of the Earth's orbit as shown.

In the left portion of the small horizontal yellow box in the lower portion of the larger light gray box near the top of Fig. 13 we show Newton's equation for centrifugal acceleration: **Ca = v²/r**. By now substituting in the orbital velocity of the Earth and the radius of the Earth's orbit into this equation, we now obtain: **Ca = .006 meters per second²** as shown by the small black box within the yellow one.

Directly below in the left-hand portion of the top light blue box we have the equation of General Relativity describing Time Dilation due to Gravity in terms of both mass and radius. We find that if we substitute in the mass of the Sun and the radius of the Earth's orbit into this equation, we now obtain the answer as shown in the small white box for the Gravitational Time Dilation, as calculated by General Relativity where: **Δt$_{G(SUN)}$ = .999,999,990,132,45**.

However, as stated in the orange box directly to the right we find that there is absolutely no way possible here for us to correctly calculate the Gravitational Acceleration for the Sun using this Gravitational Time Dilation equation of General Relativity Theory that gives us an answer that is exactly equal to the Centrifugal Acceleration that we just calculated using Newton's equation for Centrifugal Acceleration as shown

Directly below, in the left-hand portion of the larger horizontal blue box we have the equation of Action Gravity of Reality Physics that allows us to correctly calculate the Gravitational Time Dilation of the Sun at this: 1.496x10^{11} meter distance, which equals the radius of the Earth's orbit. Substituting this radius and the mass of the Sun into this equation we now obtain the Gravitational Time Dilation, as shown in the small white box: **Δt$_{G(SUN)}$ = .999,999,999,99**. We now substitute this Time Dilation into the equation as shown directly to the right to obtain the correct <u>Gravitational Acceleration</u> of the Sun at the distance of the Earth's orbit to be: **g$_{(SUN)}$ = .006 meters per second²**, as shown in the right-hand portion of the light blue box.

As we can all observe here, this result **exactly matches** the answer we obtained from Newton's equation for centrifugal force, which now completely proves the validity of this Principle of Absolute Equivalence of Reality Physics, while also, simultaneously, totally disproving the validity of the incorrect equation as proposed here by General Relativity. So, even though we sometimes hear these news reports where Theoretical Physicists claim they've performed an experiment that appears to prove General Relativity Theory, we know the are completely <u>wrong</u> because the equation they use for calculating Time Dilation due to gravity, is completely **WRONG**!

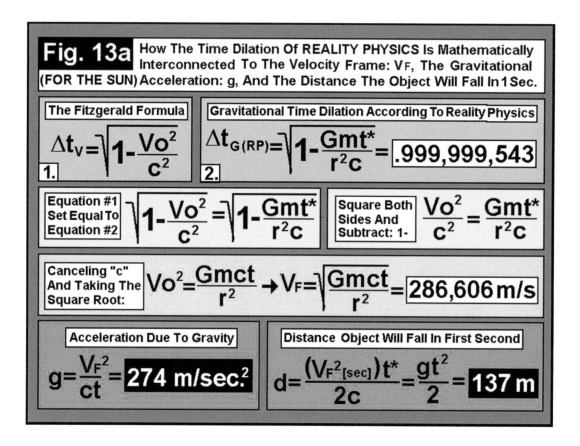

Fig. 13a How The Time Dilation Of REALITY PHYSICS Is Mathematically Interconnected To The Velocity Frame: V_F, The Gravitational (FOR THE SUN) Acceleration: g, And The Distance The Object Will Fall In 1 Sec.

The Fitzgerald Formula

$$1. \quad \Delta t_V = \sqrt{1 - \frac{V_o^2}{c^2}}$$

Gravitational Time Dilation According To Reality Physics

$$2. \quad \Delta t_{G(RP)} = \sqrt{1 - \frac{Gmt^*}{r^2 c}} = .999,999,543$$

Equation #1 Set Equal To Equation #2

$$\sqrt{1 - \frac{V_o^2}{c^2}} = \sqrt{1 - \frac{Gmt^*}{r^2 c}}$$

Square Both Sides And Subtract: 1-

$$\frac{V_o^2}{c^2} = \frac{Gmt^*}{r^2 c}$$

Canceling "c" And Taking The Square Root:

$$V_o^2 = \frac{Gmct}{r^2} \rightarrow V_F = \sqrt{\frac{Gmct}{r^2}} = 286,606 \text{ m/s}$$

Acceleration Due To Gravity

$$g = \frac{V_F^2}{ct} = 274 \text{ m/sec.}^2$$

Distance Object Will Fall In First Second

$$d = \frac{(V_F{}^2{}_{[sec]})t^*}{2c} = \frac{gt^2}{2} = 137 \text{ m}$$

We can further extend and illustrate how Reality Physics reveals the complete disproof of General Relativity by now observing Fig. 13a above. We simply set equation #1 equal to equation # 2 to obtain the equation in the left yellow box directly below. We now square both sides and subtract out the: "**1-'s**" to obtain the equation in the right yellow box. We now, as shown in the light blue box, cancel out the: "**c's**", and then take the square root of both sides to arrive at the equation in the middle of the light blue box.

We now substitute in both the mass and radius of the Sun into this equation to obtain values that exactly match, and are able to explain, the results we obtain from the equations of Newtonian Classical Physics. What we are doing here is revealing the actual fundamental physics that shows how the mathematical amount of gravitational acceleration, as calculated by the equations of Newtonian Classical Dynamics, is explained by the Active Geometry of Action Gravity. This now completely proves that Reality Physics is correct because we can obtain these correct answers that have existed in Newtonian Physics for over three and a half centuries now.

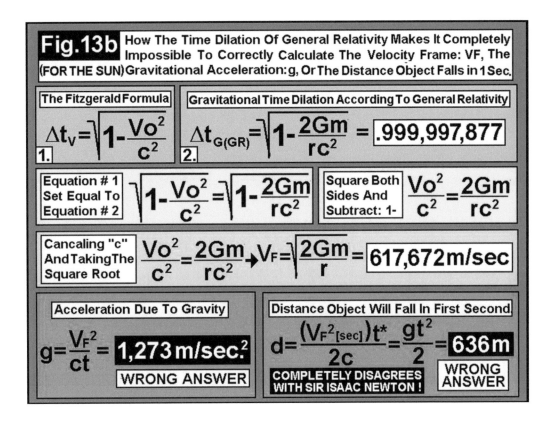

Fig. 13b How The Time Dilation Of General Relativity Makes It Completely Impossible To Correctly Calculate The Velocity Frame: VF, The (FOR THE SUN) Gravitational Acceleration: g, Or The Distance Object Falls in 1 Sec.

The Fitzgerald Formula

$$\Delta t_V = \sqrt{1 - \frac{Vo^2}{c^2}}$$

1.

Gravitational Time Dilation According To General Relativity

$$\Delta t_{G(GR)} = \sqrt{1 - \frac{2Gm}{rc^2}} = .999,997,877$$

2.

Equation # 1 Set Equal To Equation # 2

$$\sqrt{1 - \frac{Vo^2}{c^2}} = \sqrt{1 - \frac{2Gm}{rc^2}}$$

Square Both Sides And Subtract: 1-

$$\frac{Vo^2}{c^2} = \frac{2Gm}{rc^2}$$

Cancaling "c" And Taking The Square Root

$$\frac{Vo^2}{c^2} = \frac{2Gm}{rc^2} \rightarrow V_F = \sqrt{\frac{2Gm}{r}} = 617,672 \text{ m/sec}$$

Acceleration Due To Gravity

$$g = \frac{V_F^2}{ct} = 1,273 \text{ m/sec}^2$$

WRONG ANSWER

Distance Object Will Fall In First Second

$$d = \frac{(V_F^2{}_{[sec]})t^*}{2c} = \frac{gt^2}{2} = 636 \text{ m}$$

COMPLETELY DISAGREES WITH SIR ISAAC NEWTON !

WRONG ANSWER

Now, turning our attention to Fig. 13b, we observe how the General Theory of Relativity describes gravity at the Sun's Surface. Again, setting these two equations equal, we perform the same operation as we did before in Fig. 13a to now calculate a Velocity Frame by the equations of General Relativity: **VF = 617,672 meters per second**.

This is more than twice the correct value we calculated before using the equations of Action Gravity of Reality Physics which gave us a: **VF = 286,606 meters per second**, as is illustrated in the large light blue horizontal box back in Fig. 13a.

Consequently, we can now understand why the values we obtain from General Relativity give us the <u>incorrect</u> answer for both the gravitational acceleration: "g", and the distance an object will fall in one second at the Sun's surface: "d". This is shown by the black boxes located at the very bottom of the of Fig. 13b as shown. This proves, beyond any possible doubt what so ever, that Einstein's equations of General Relativity Theory will <u>NEVER</u> agree with the equation of Newtonian Classical Dynamics.

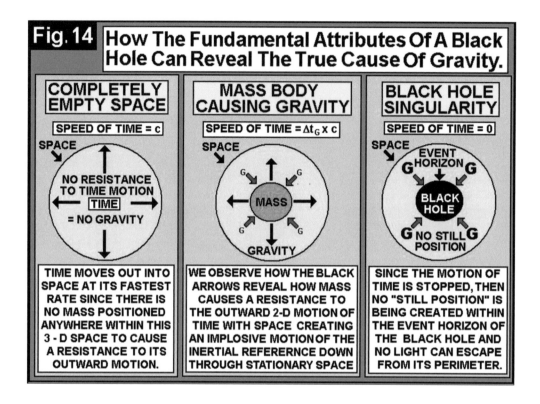

Fig. 14 How The Fundamental Attributes Of A Black Hole Can Reveal The True Cause Of Gravity.

COMPLETELY EMPTY SPACE

SPEED OF TIME = c

SPACE

NO RESISTANCE TO TIME MOTION
TIME
= NO GRAVITY

TIME MOVES OUT INTO SPACE AT ITS FASTEST RATE SINCE THERE IS NO MASS POSITIONED ANYWHERE WITHIN THIS 3 - D SPACE TO CAUSE A RESISTANCE TO ITS OUTWARD MOTION.

MASS BODY CAUSING GRAVITY

SPEED OF TIME = Δt_G x c

SPACE

G G
G MASS G
G G
GRAVITY

WE OBSERVE HOW THE BLACK ARROWS REVEAL HOW MASS CAUSES A RESISTANCE TO THE OUTWARD 2-D MOTION OF TIME WITH SPACE CREATING AN IMPLOSIVE MOTION OF THE INERTIAL REFERERNCE DOWN THROUGH STATIONARY SPACE

BLACK HOLE SINGULARITY

SPEED OF TIME = 0

SPACE
EVENT HORIZON
G G
BLACK HOLE
G NO STILL G
POSITION

SINCE THE MOTION OF TIME IS STOPPED, THEN NO "STILL POSITION" IS BEING CREATED WITHIN THE EVENT HORIZON OF THE BLACK HOLE AND NO LIGHT CAN ESCAPE FROM ITS PERIMETER.

Now, turning our attention to the left diagram in Fig. 14, we show how the Now Point in Time passes outward into space at its greatest rate when space is empty of matter. In the middle diagram we show how a body of mass causes a RESISTANCE to this outward motion of time with space causing a vicinity of: less active pressure density at the location of the body of mass now producing the gravitational acceleration it must experience. In the right diagram we show what happens when we accumulate so much mass that we completely stop this outward Motion of Time with space, now meaning that, no more "**stationary position**" within the universe can be formed here because this mass, which has now condensed into a Black Hole, is preventing time from passing outward into space.

What this proves is that it takes the continual outward Motion of Time with Space at "**c**" to form the stationary position, that Reality Physics calls: "**The Absolute Universal Metric**" within which the entire universe exists. If you stop this continual motion of time passing outward into space, you also immediately stop the continual creation of the stationary position being formed in the universe and form a "Black Hole" where time can no longer pass out into space.

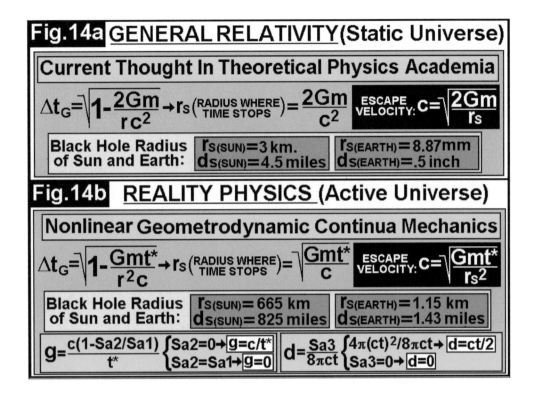

In the light gray box near the top of Fig. 14a we show the formula that Karl Schwarzschild used to calculate the radius of a Black Hole using the equation of General Relativity. In the orange boxes we show the diameter: **ds**, and the radius: **rs**, of the Sun and Earth if their entire mass was reduced down to the size of a Black Hole.

In the light gray box in Fig. 14b, just below it, we have the formula from an entirely new study recently developed in Reality Physics called: "Nonlinear Geometrodynamic Continua Mechanics", which now reveals this new equation of Action Gravity we use to calculate the: "**rs**", or the "**radius at which time stops**". In the orange boxes below, we have the Block Hole radii, and the diameters of both the Sun and the Earth reduced down to the size of a Black Hole. In the black boxes to the right, we show the escape velocity of the Black Hole to be: "**c**" in both cases. In the bottom of Fig. 14b we show how Active Geometry calculates a: **g=c/t*** (an acceleration) when **Sa2=0**, and a: **g=0**, when **Sa2=SA1**. Also, we show how when **Sa3=Sa1** or: **4π(ct)²**, the distance the object will fall will be: **d=.5ct**, in one second at the Black Holes surface and when **Sa3=0**, then **d=0**

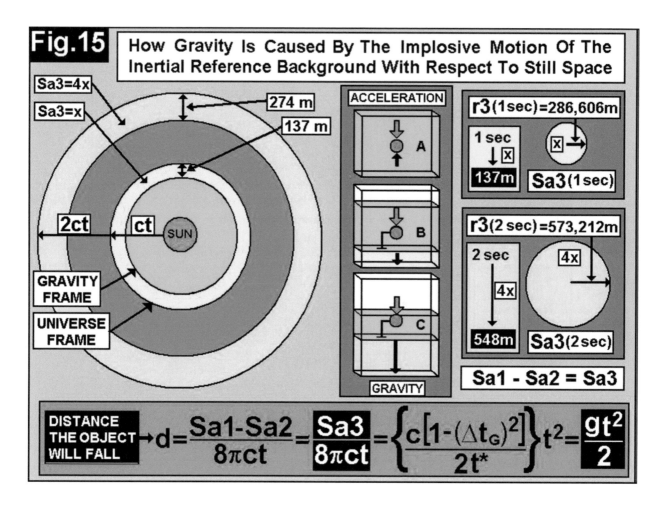

Fig. 15 How Gravity Is Caused By The Implosive Motion Of The Inertial Reference Background With Respect To Still Space

$$d = \frac{Sa1 - Sa2}{8\pi ct} = \frac{Sa3}{8\pi ct} = \left\{ \frac{c\left[1-(\Delta t_G)^2\right]}{2t^*} \right\}t^2 = \frac{gt^2}{2}$$

In the left portion of Fig. 15 we have an Active Geometry Diagram where we illustrate how a body of mass, like the Sun for example, causes a resistance to the active, two-dimensional, outward motion of time with spacetime with space. This is indicated by the colored concentric circles surrounding the body of mass at their center.

From this diagram, and from the two small details in the far-right side of Fig. 15, we immediately observe how both of these surface area threes, or "**Sa3's**", as represented by the yellow and light blue colored circles, would increase in size at the nonlinear rate at which the surface area of a sphere increases in size with respect to a linear increase in its radius, as time continues to pass out into space a "**c**".

This now, for the very first time in the history of physics reveals to us the real reason as to why all small solid objects in a gravitational field fall <u>four</u> times further after <u>two seconds</u>, <u>nine</u> times further after <u>three seconds</u>, etc., than after one second.

This is because they always follow the nonlinear rate at which these two surface areas of Time Displacement, surface area one, or **Sa1**, representing the outward Time Displacement within the universe frame, and the smaller surface area two: or **Sa2**, representing the active Displacement of Time in the Gravity Frame, as both now continue to expand outward from their center point, as shown.

As illustrated in the light gray box in the center of Fig. 15, we show how this ability of a body of mass to slow the outward motion of time is what actually causes the stationary position of the inertial reference background of the universe to accelerate downward with an implosive motion towards the surface of the body of mass. This is indicated by the vertical black arrows pointing downward in the two small details in the bottom of the vertical gray box.

From this we can now understand how the Principle of Absolute Equivalence illustrates to us how this downward acceleration of the inertial reference background, as indicated here by the yellow and blue rectangles, produces a gravitational force, as indicated by the small orange arrows pointing downward in the diagram.

As shown by the top detail in this vertical gray box, this exact same force can also be produced by accelerating an object with respect to the still position of the Absolute Universal Metric (green square box) of the universe where both the still position of space, and the stationary position in the inertial reference background, are three-dimensionally aligned. The only difference is that: **<u>with gravity the stationary position of the inertial reference background of the universe accelerates downward with respect to space creating the gravitational force, as indicated by the small orange arrows.</u>**

<u>Gravity is an action because it results from an action</u>. Like a leaf that immediately starts moving when it hits the surface of a running stream, a small stone immediately falls downward when it's released because it now occupies a stationary position within the inertial reference background of the universe that is **<u>already moving</u>** down through the still position of space at the surface of the body of mass, carrying the small stone down along with it.

ACTION GRAVITY

THE TRUE CAUSE OF GRAVITY REVEALED
by JEFF LEE CENTER FOR REALITY PHYSICS

From what we have discovered here, we now understand that in order for gravity to exist we, first of all, need a continual outward motion of time passing out into space to create an inertial reference background pressure density. By placing a large body of mass into this new active inertial reference background pressure density, we slow down this outward motion of time to now create a vicinity of less active pressure density where the mass is located. This is what causes the implosive accelerating motion of the inertial reference down through space at the surface of the body of mass. However, because General Relativity doesn't even know about this new discovery of the: "Speed of Time Through Space", they could try for the next twenty-thousand years to understand the cause of gravity, and they would be no closer than they are now since gravity is an action, not a warping or curvature of space.

As we all know, since the General Theory of Relativity incorrectly assumes a static universe, there is absolutely no possible way they will ever be able to discover the true cause of gravity. This is why they all continue to describe gravity as "curved, or warped, space, or spacetime". So how does space curve, or warp, anyway? Do certain particles of space get closer together in one given vicinity of space, and then further apart in another vicinity? Are there compressions and rarefactions in space just like in the air?

Obviously, as we can readily observe here, the real reason for this problem is because Theoretical Physics Academia is in the wrong universe. This is why they could try forever to use the General Theory of Relativity to reveal the true cause of gravity, and they would find themselves to be no closer to accomplishing this than they are right now since there is simply no way they can ever create a working universe that has an immediate downward acceleration of the still position of the inertial reference background down through space at the surface of a large body of mass using the static universe of General Relativity.

Using the Action Gravity of Reality Physics, we have revealed that gravity is an "instantaneous downward acceleration" that causes all small solid objects released in a gravitational field to immediately fall down towards a large body of mass. Why? Because it "moves along" with the still position of the inertial reference background as it now implodes with an accelerating motion down through space at the surface of the body of mass.

What this means is that the action necessary to transport the small object to the ground is already in place before the small object is released. This is the same as the example we gave when we dropped a leaf into the sunning stream. Because the running stream is already in motion the leaf immediately begins to move as soon as it makes contact with the running water, just like the falling stone does in an "active" universe.

What we have to realize here is that the still position within the inertial reference background, just like the still position within the running stream, is already in motion down through space. This is why a small object instantaneously accelerates downward in a gravitational field as soon as you let it loose from you hand. This is also why, as revealed here by this new Principle of Absolute Equivalence, a small object located on the surface of a large body of mass will experience a force even though it is not moving through space. It is then obvious that this downward force of gravity could only be produced by letting the still position of the inertial reference background implode down through space, at the surface of the large mass.

As was revealed here by both the Active Geometry Diagrams of Reality Physics, and also the equations of Action Gravity, gravity is an "**action**", and NOT just a "warpage, or curvature of space, or spacetime". It will be interesting to see how long it takes Theoretical Physics Academia to figure this out, but we hope it's soon so we can all know the true cause of gravity.

Printed in the United States
by Baker & Taylor Publisher Services